AF388897

Natural Products from Marine Fungi

Natural Products from Marine Fungi

Editor

Hee Jae Shin

MDPI • Basel • Beijing • Wuhan • Barcelona • Belgrade • Manchester • Tokyo • Cluj • Tianjin

Editor
Hee Jae Shin
Marine Natural Products Chemistry Laboratory
Korea Institute of Ocean Science & Technology
Korea

Editorial Office
MDPI
St. Alban-Anlage 66
4052 Basel, Switzerland

This is a reprint of articles from the Special Issue published online in the open access journal *Marine Drugs* (ISSN 1660-3397) (available at: https://www.mdpi.com/journal/marinedrugs/special_issues/fungi).

For citation purposes, cite each article independently as indicated on the article page online and as indicated below:

LastName, A.A.; LastName, B.B.; LastName, C.C. Article Title. *Journal Name* **Year**, *Article Number*, Page Range.

ISBN 978-3-03936-772-6 (Hbk)
ISBN 978-3-03936-773-3 (PDF)

Contents

About the Editor

Hee Jae Shin is a principal research Scientist at the Marine Biotechnology Center, Korea Institute of Ocean Science and Technology (KIOST) and a professor at the University of Science and Technology. He received his PhD from the University of Tokyo (1997), where he studied the isolation and structure determination of protease inhibitors from cyanobacteria. He undertook two post-doctoral positions at the Marine Biotechnology Institute, Japan (1997–1999) and the center for marine biotechnology and biomedicine, Scripps Institution of Oceanography with Professor William Fenical (1999–2000). He spent 3 years working in the pharmaceutical industry (2000–2003) and then returned to the Korea Ocean Research and Development Institute, which is now KIOST. His research interest is on the isolation and structure determination of bioactive marine natural products from marine microorganisms including fungi, actinomycetes, deep-sea and symbiotic microorganisms, and discovery and development of drug candidates.

Editorial

Natural Products from Marine Fungi

Hee Jae Shin [1,2]

1   Marine Natural Products Chemistry Laboratory, Korea Institute of Ocean Science and Technology, 385 Haeyang-ro, Yeongdo-gu, Busan 49111, Korea; shinhj@kiost.ac.kr
2   Department of Marine Biotechnology, University of Science and Technology (UST), 217 Gajungro, Yuseong-gu, Daejeon 34113, Korea

Received: 20 April 2020; Accepted: 25 April 2020; Published: 27 April 2020

Introduction

Marine fungi have been studied since the first record of the species *Sphaeria posidoniae* (*Halotthia posidoniae*) on the rhizome of the sea grass *Posidonia oceanica* by Durieu and Montagne in 1846 [1], but they have largely been neglected, even though it is estimated that there are greater than 10,000 marine fungal species. To date, a relatively small percentage of described species are associated with marine environments, with ~1100 species exclusively retrieved from the marine environment, although estimates for the number of fungal species on the planet range from 1.5 to over five million, likely fewer than 10% of fungi have been identified so far. Fungi have been found in nearly every marine habitat examined, including sediments, the water column, driftwood, sessile and mobile invertebrates, algae, and marine mammals, ranging in location from the deep sea all the way to surface waters [2].

It is believed that the exploration of marine fungi that are living in new and extreme habitats will advance the isolation of novel marine fungi and, thus, might lead to the isolation of novel secondary metabolites.

Marine Fungal Metabolites

When considering that 38% of the approximately 22,000 bioactive microbial metabolites are of fungal origin, and that only about 5% of the world's fungal taxa have been described, fungi exhibit a tremendous potential for the discovery of novel bioactive secondary metabolites [3]. Specifically, marine fungi still represent an underestimated but rich source for new secondary metabolites, although their distribution and ecological role often remain scarce. Marine fungi are an important source of secondary metabolites useful for the drug discovery. Even though marine fungi are less explored in comparison to their terrestrial counterparts, a number of useful hits have been obtained from the drug discovery perspective adding to their importance in the natural product discovery [4]. A number of metabolites from marine fungi have been discovered from various sources, which show a range of activities, such as antibacterial, antiviral, and anticancer agents. Although, over a thousand marine fungi related secondary metabolites have already been reported, except for cyclosporine A, none of them have reached the market yet, which could partly be related to non-comprehensive screening approaches and a lack of sustained lead optimization. Marine fungi are potent producers of polyketides, alkaloids, terpenes, peptides, and mixed biosynthesis compounds that represent chemical groups of secondary metabolites.

In the review article by Zhao et al. (2018) [5], the novel natural products from extremophilic fungi were compiled. The authors focused on 314 novel compounds from 56 extremophilic fungal strains published from 2005 to 2017, highlighting the chemical structures and their biological potential. Extremophilic fungi have been found to develop unique defenses to survive extremes of pressure, temperature, salinity, desiccation, and pH, leading to the biosynthesis of novel natural products with diverse biological activities. This review focuses on the source, chemical structure types, biological activities, and references of all novel natural products and will help readers to better understand the underlying potential of fungal natural products as drug candidates. This review demonstrated that fungi from extreme environments are a rich source for novel natural products, even though the research on them is not as up-to-date as the research on fungi in other mesophilic environments due to the difficulties in both sample collection and cultivation.

Deep-sea fungi inhabit depths of thousand meters or below the surface where the sea environments are extreme, being typically characterized by the absence of sunlight irradiation, predominantly low temperature, high hydrostatic pressure, and oligotrophy [6]. Since the first report of fungal isolation from deep-sea [7], many fungi have been isolated from various deep-sea environments. Fungi isolated from the deep-sea samples are one of the most pivotal and promising source for bioactive compounds, presumably owing to the chemical diversity and biodiversity of their secondary metabolites that could be used for drug discovery and pharmacological applications [8].

From the deep-sea fungus *Phomopsis lithocarpus* isolated from a deep-sea sediment sample collected in the Indian Ocean at the depth of 3606 m, Zhang et al. isolated five new benzophenone derivatives, sharing a rare naturally occurring aldehyde functionality in this family, and a new eremophilane derivative, together with two known compounds [8]. One of the new compounds, tenellone H, exhibited cytotoxic activity against HepG-2 and A549 cell lines with IC_{50} values of 16.0 and 17.6 µM, respectively.

From another deep-sea fungal strain FL30r, identified as *Phialocephala* sp. obtained from a deep-sea sample (depth 5059 m), two new nitrogen-containing sorbicillinoids, named sorbicillasins A and B, and a new 3,4,6-trisubstituted α-pyrone derivative, scirpyrone K, together with two known biosynthetically related polyketides, were isolated by the OSMAC (one strain-many compounds) method [9]. Sorbicillasins A and B were unusual naturally occurring nitrogen-containing sorbicillinoid derivatives with a novel hexa-hydropyrimido[2, 1-*a*]isoindole moiety. Scirpyrone K exhibited radical scavenging activity against DPPH.

Symbiotic relationships are vast and diverse within the marine environment and many marine organisms, such as invertebrates as well as other marine macro-organisms, live in symbiosis with their microbial communities [10,11]. Marine sponges, sometimes referred to as microbial fermenters, are an outstanding source of highly diverse microbial communities, including new fungal species [12]. Sponge-derived fungi are one of the richest sources of many structurally unique and biologically active secondary metabolites among marine sources [13]. Seven new structurally diverse polyketide derivatives, along with 21 known compounds, were isolated from cultures of the sponge-derived fungus, *Alternaria* sp. SCSIO41014 by Liu et al. [14]. Altertoxin VII exhibited cytotoxic activities against human erythroleukemia (K562), human gastric carcinoma cells (SGC-7901), and hepatocellular carcinoma cells (BEL-7402), with IC_{50} values of 26.58 ± 0.80, 8.75 ± 0.13, and 13.11 ± 0.95 µg/mL, respectively. This compound is the first example possessing a novel 4,8-dihydroxy-substituted perylenequinone derivative, while the phenolic hydroxy groups have always commonly substituted at C-4 and C-9. In the work of Zu et al. [15], two novel aspochalasins, tricochalasin A, and aspochalasin A2, along with three known compounds, have been isolated from the different culture broth of *Aspergillus* sp., which was found in the gut of a marine isopod *Ligia oceanica* by employing the OSMAC approach by varying the culture conditions. Eight new fungal natural products, meroterpenoids and isocoumarinoids, were isolated from the culture of the salt-tolerant plant-associated fungus *Myrothecium* sp. OUCMDZ-2784 [16]. This study revealed that fungi living in the salt-tolerant plants are important biological resources for new and bioactive natural products.

Most of the studies on marine fungi have been made with those that are associated with marine sediments taken from shallow water or deep-sea and mangrove areas [17]. Five new anthraquinone derivatives, auxarthrols D–H, along with two known analogues, were isolated from the culture of the marine sediment-derived fungus *Sporendonema casei* [18]. Auxarthrols D and F showed cytotoxic activities, with IC_{50} values from 4.5 µM to 22.9 µM, while altersolanol B displayed potential antitubercular activity for the first time. In the experimental work of Song et al., six new diketopiperazines, three pairs of new brevianamides, were isolated from a marine-derived fungus strain *Aspergillus versicolor* MF180151 that was isolated from a sediment sample [19]. Shin et al. isolated six new phenalenone derivatives and five known compounds of the herqueinone class from a marine sediment-derived fungus *Penicillium* sp. [20]. 4-Hydroxysclerodin and an acetone adduct of a triketone exhibited moderate anti-angiogenetic and anti-inflammatory activities, respectively. A new alkenoic acid, fusaridioic acid A, three new bis-alkenoic acid esters, fusariumester A1, A2, and B, together with three known compounds, were isolated from the fungus *Fusarium solani* H915 derived from mangrove sediments by Qiu et al. [21]. Hymeglusin, an alkenoic acid derivative with a β-lactone ring, showed potent antifungal activity against tea pathogenic fungi with low toxicity.

In summary, the special issue "Natural Products from Marine Fungi" compiles the recent results from marine fungi. As a guest editor, I am grateful to all the authors who contributed their excellent results to the special issue, all the reviewers who carefully evaluated the submitted manuscripts, and the editorial boards of Marine Drugs, and Vincent Di, Assistant Editor, for their support and kind help.

Funding: This research was supported in part by the Korea Institute of Ocean Science and Technology (Grant PE99852).

Conflicts of Interest: The authors declare no conflict of interest.

References

1. Montagne, C. *Exploration scientifique de l'Algérie pendant les années 1840, 1841, 1842*; Durieu De Maisonneuve, M.C., Ed.; Imprimerie Royale: Paris, France, 1846; pp. 1–197.

2. Amend, A.; Burgaud, G.; Cunliffe, M.; Edgcomb, V.P.; Ettinger, C.L.; Gutierrez, M.H.; Heitman, J.; Hom, E.F.Y.; Ianiri, G.; Jones, A.C.; et al. Fungi in the marine environment: Open questions and unsolved problems. *MBio* **2019**, *10*, e01189-18. [CrossRef] [PubMed]

3. Schulz, B.; Draeger, S.; de la Cruz, T.E.; Rheinheimer, J.; Siems, K.; Loesgen, S.; Bitzer, J.; Schloerke, O.; Zeeck, A.; Kock, I.; et al. Screening strategies for obtaining novel, biologically active, fungal secondary metabolites from marine habitats. *Bot. Mar.* **2008**, *51*, 219–234. [CrossRef]

4. Butler, M.S.; Robertson, A.A.; Cooper, M.A. Natural product and natural product derived drugs in clinical trials. *Nat. Prod. Rep.* **2014**, *31*, 1612–1661. [CrossRef] [PubMed]

5. Zhang, X.; Li, S.-J.; Liang, Z.-Z.; Zhao, C.-Q. Novel natural products from extremophilic fungi. *Mar. Drugs* **2018**, *16*, 194–230. [CrossRef] [PubMed]

6. Deshmukh, S.K.; Prakash, V.; Ranjan, N. Marine fungi: A source of potential anticancer compounds. *Front. Microbiol.* **2018**, *8*, 2536–2560. [CrossRef] [PubMed]

7. Roth, F.J.; Orpurt, P.A.; Ahearn, D.J. Occurrence and distribution of fungi in a subtropical marine environment. *Can. J. Bot.* **1964**, *42*, 375–383. [CrossRef]

8. Xu, J.L.; Liu, H.-X.; Chen, Y.-C.; Tan, H.-B.; Guo, H.; Xu, L.-Q.; Li, S.-N.; Huang, Z.-L.; Li, H.-H.; Gao, X.-X.; et al. Highly substituted benzophenone aldehydes and eremophilane derivatives from the deep-sea derived fungus *Phomopsis lithocarpus* FS508. *Mar. Drugs* **2018**, *16*, 329. [CrossRef] [PubMed]

9. Zhang, Z.; He, X.; Che, Q.; Zhang, G.; Zhu, T.; Gu, Q.; Li, D. Sorbicillasins A–B and Scirpyrone K from a Deep-Sea-Derived Fungus, Phialocephala sp. FL30r. *Mar. Drugs* **2018**, *16*, 245. [CrossRef] [PubMed]

10. Webster, N.S.; Taylor, M.W. Marine sponges and their microbial symbionts: Love and other relationships. *Environ. Microbiol.* **2012**, *14*, 335. [CrossRef] [PubMed]

11. Armstrong, E.; Yan, L.; Boyd, K.G.; Wright, P.C.; Burgess, J.G. The symbiotic role of marine microbes on living surfaces. *Hydrobiologia* **2001**, *461*, 37. [CrossRef]

12. Bovio, E.; Garzoli, L.; Poli, A.; Prigione, V.; Firsova, D.; McCormack, G.P.; Varese, G.C. The culturable mycobiota associated with three Atlantic sponges, including two new species: *Thelebolus balaustiformis* and *T. spongiae*. *Fungal Syst. Evol.* **2018**, *1*, 141. [CrossRef]

13. Bugni, T.S.; Ireland, C.M. Marine-derived fungi: A chemically and biologically diverse group of microorganisms. *Nat. Prod. Rep.* **2004**, *21*, 143. [CrossRef] [PubMed]

14. Pang, X.; Lin, X.; Wang, P.; Zhou, X.; Yang, B.; Wang, J.; Liu, Y. Perylenequione derivatives with anticancer activities isolated from the marine sponge-derived fungus, *Alternaria* sp. SCSIO41014. *Mar. Drugs* **2018**, *16*, 280. [CrossRef] [PubMed]

15. Li, X.; Ding, W.; Wang, P.; Xu, J. Two novel aspochalasins from the gut fungus *Aspergillus* sp. Z4. *Mar. Drugs* **2018**, *16*, 343. [CrossRef] [PubMed]
16. Xu, Y.; Wang, C.; Liu, H.; Zhu, G.; Fu, P.; Wang, L.; Zhu, W. Meroterpenoids and isocoumarinoids from a *Myrothecium* fungus associated with *Apocynum venetum*. *Mar. Drugs* **2018**, *16*, 363. [CrossRef] [PubMed]
17. Imhoff, J.F. Natural products from marine fungi-still an underrepresented resource. *Mar. Drugs* **2016**, *14*, 19. [CrossRef] [PubMed]
18. Ge, X.; Sun, C.; Feng, Y.; Wang, L.; Peng, J.; Che, Q.; Gu, Q.; Zhu, T.; Li, D.; Zhang, G. Anthraquinone derivatives from a marine-derived fungus *Sporendonema casei* HDN16-802. *Mar. Drugs* **2019**, *17*, 334. [CrossRef] [PubMed]
19. Hu, J.; Li, Z.; Gao, J.; He, H.; Dai, H.; Xia, X.; Liu, C.; Zhang, L.; Song, F. New diketopiperazines from a marine-derived fungus strain *Aspergillus versicolor* MF180151. *Mar. Drugs* **2019**, *17*, 262. [CrossRef] [PubMed]
20. Park, S.C.; Julianti, E.; Ahn, S.; Kim, D.; Lee, S.K.; Noh, M.; Oh, D.-C.; Oh, K.-B.; Shin, J. Phenalenones from a marine-derived fungus *Penicillium* sp. *Mar. Drugs* **2019**, *17*, 176. [CrossRef] [PubMed]
21. Liu, S.-Z.; Yan, X.; Tang, X.-X.; Lin, J.-G.; Qiu, Y.-K. New bis-alkenoic acid derivatives from a marine-derived fungus *Fusarium solani* H915. *Mar. Drugs* **2018**, *16*, 483. [CrossRef] [PubMed]

Review

Novel Natural Products from Extremophilic Fungi

Xuan Zhang, Shou-Jie Li, Jin-Jie Li, Zi-Zhen Liang and Chang-Qi Zhao *

Gene Engineering and Biotechnology Beijing Key Laboratory, Key Laboratory of Cell Proliferation and Regulation Biology, Ministry of Education, College of Life Science, Beijing Normal University, Beijing 100875, China; zhangxuan28@outlook.com (X.Z.); LSJ19930801@163.com (S.-J.L.); lijinjie.7785004@163.com (J.-J.L.); zizhenliang@gmail.com (Z.-Z.L.)

* Correspondence: 04020@bnu.edu.cn; Tel.: +86-5880-5046; Fax: +86-10-5880-7720

Received: 3 May 2018; Accepted: 2 June 2018; Published: 4 June 2018

Abstract: Extremophilic fungi have been found to develop unique defences to survive extremes of pressure, temperature, salinity, desiccation, and pH, leading to the biosynthesis of novel natural products with diverse biological activities. The present review focuses on new extremophilic fungal natural products published from 2005 to 2017, highlighting the chemical structures and their biological potential.

Keywords: natural products; extremophilic fungi; biological activity

1. Introduction

The term "extremophile" was first proposed by MacElroy in 1974 to describe a broad group of organisms which lived optimally under extreme conditions [1], and the taxonomic range of them has been expanded from prokaryotes to all three domains—Eucarya, Bacteria, and Archaea [2]. Extremophiles are classified into seven categories according to different extreme habitats. Piezophiles reside under high hydrostatic pressure, which have been isolated from the deep-sea sediments (>3000 m depth) and the guts of bottom-dwelling animals [3–5]. Organisms whose optimal growth temperature ranges from 50 to 80 °C or exceeds 80 °C are called thermophiles or hyperthermophiles respectively, which have been mainly isolated from hot springs, deep-sea hydrothermal vents, and decaying plant matter [6]. Psychrophiles living in the other extreme thermal habitat have been obtained from the Antarctic, the Arctic, and glacial regions [7]. Halophiles are defined as organisms requiring >3% NaCl for growth [6]. Xerophiles thrive under the desiccated and have been discovered in ashes and deserts [8]. Acidophiles or alkaliphiles show optimal growth at pH values <4 and >9 respectively [6]. Organisms from these extreme habitats require special survival strategies for growing and reproducing, and the adaptation to such conditions requires the modification of gene regulation and metabolic pathways [9–13], thus extremophiles seem to be good potential candidates for novel natural products.

Several reviews discussing natural products from special environments have been published over the last decade, and the topics include natural products from cold water [14,15], polar regions [16,17], and deep sea [18,19]. In 2009 Wilson and Brimble reviewed the studies on the structure of molecules isolated from the extremes of life [6], while those compounds were mainly isolated from bacteria and actinomycetes. Few reviews focused on the secondary metabolites from extremophilic fungi.

This review focuses on the source, chemistry, and biology of novel natural products which were derived from extremophilic fungi. These fungal products are classified according to extremophile classifications, and when a fungal strain falls under multiple classifications, it is grouped under the dominant environmental factor. In addition, the table (Table 1) including the chemical structure types, biological activities and references of all novel natural products will help readers to better understand their underlying potential as drug candidates. It is worth noting that in compiling this review all isolated strains were selected strictly according to the above categories and that strains which do not meet the standards were not cited. For example, one strain was isolated from the deep sea but its underwater depth was less than 3000 m, which did not meet the criteria of piezophiles. Therefore, we did not include it in this review.

2. Piezophilic Fungi

Phialocephala sp. FL30r obtained from an underwater sample (depth 5059 m, the east Pacific) was a powerful producer of diverse sorbicillin-type compounds. Two new bisorbicillinoids, oxosorbiquinol (**1**) and dihydrooxosorbiquinol (**2**) (Figure 1) [20], four new sorbicillin trimers, trisorbicillinones A–D (**3–6**) (Figure 1) [21,22], one new sorbicillin dimer, dihydrotrichodermolide (**7**) (Figure 1), one new sorbicillin monomer, dihydrodemethylsorbicillin (**8**) (Figure 1), and one novel benzofuran derivative, phialofurone (**9**) (Figure 1) [23] have been described from this fungal strain since 2007. The cytotoxic activity (IC$_{50}$) of compounds **1** and **2** against several cancer cell lines (P388, HL60, BEL7402, and K562) ranged from 8.9 to 68.2 µM. Compound **3** showed cytotoxic activity against P388 and HL60 cells with IC$_{50}$ values of 9.10 and 3.14 µM respectively, while compounds **4–7** exhibited weaker activities against P388 and K562 cells. Compounds **8** and **9** exhibited potent activities against P388 cells with IC$_{50}$ values of 0.1 and 0.2 µM respectively.

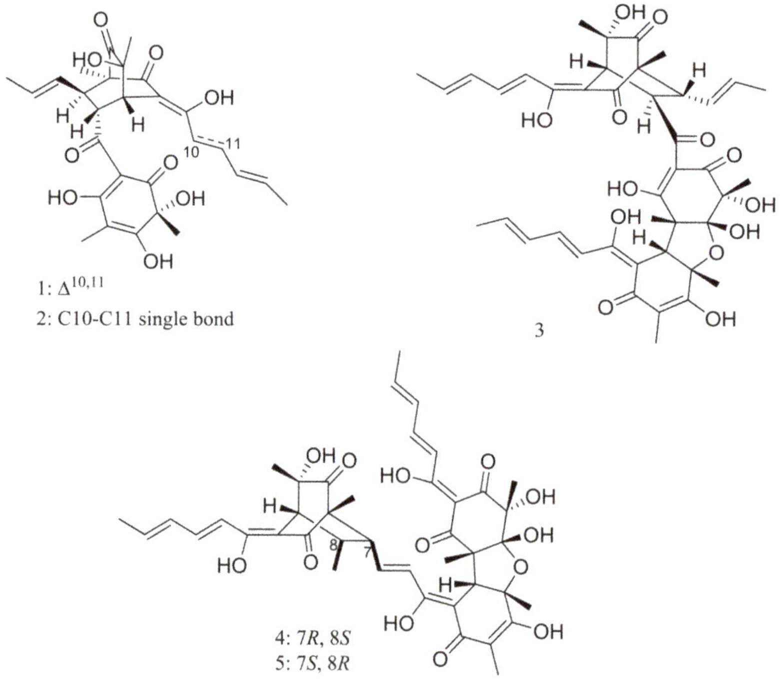

Figure 1. *Cont.*

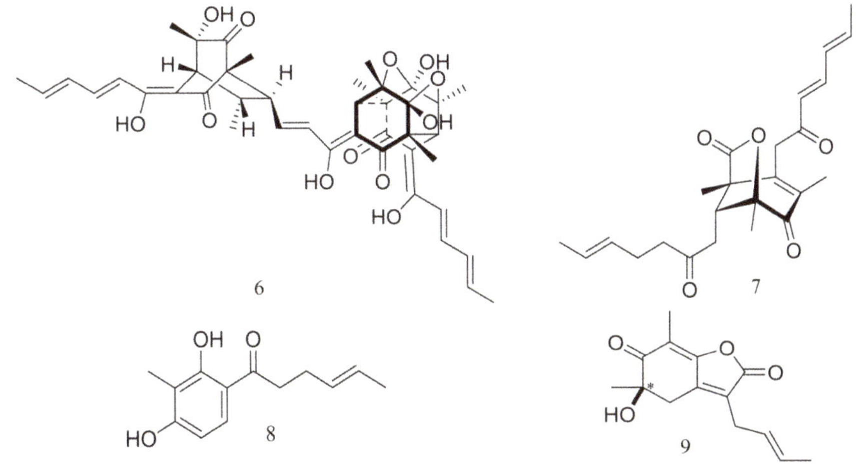

Figure 1. Novel natural products derived from piezophilic fungi (compounds **1–9**). * Absolute configuration is not determined.

Brevicompanines D–H (**10–14**) (Figure 2) are five new diketopiperazine alkaloids produced by the deep-sea sediment-derived fungus *Penicillium* sp. F1 (depth 5080 m). Compounds were assessed for their anti-inflammatory activities on LPS-challenged BV2 cells, **11** and **14** displaying IC_{50} values of 27 and 45 µg/mL respectively, other compounds being found inactive. According to the structure-bioactivity relationship, authors supposed that substitutions at the N-6 position may contribute to the anti-inflammatory activity [24].

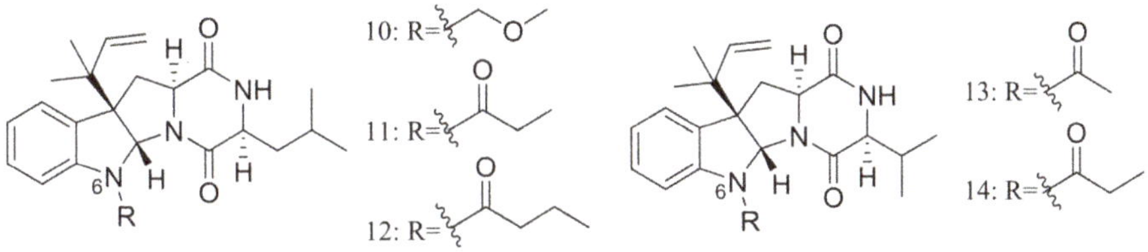

Figure 2. Novel natural products derived from piezophilic fungi (compounds **10–14**).

In 2009 and 2010 several new alkaloids named meleagrins B–E (**15–18**) and roquefortines F–I (**19–22**) (Figure 3) , along with six new diterpenes named conidiogenones B–G (**23–28**) (Figure 3) were described from the deep-sea sediment-derived *Penicillium* sp. F23-2 (depth 5080 m) [25,26]. Compound **15** showed moderate cytotoxic activity against HL60, MOLT4, A549, and BEL7402 cell lines with IC_{50} values ranging from 1.8 to 6.7 µM, while compounds **17** and **18** showed weaker activities against A549 cell line with IC_{50} values of 32.2 and 55.9 µM respectively. Further elucidation of the potential cytotoxic mechanism by flow cytometric analysis indicated that **15** could induce HL60 cells apoptosis at 5 and 10 µM. In addition, compound **24** showed potent selective cytotoxic activity against HL60 and BEL7420 cells with IC_{50} values of 0.038 and 0.97 µM respectively. This study represented the first report on the antitumor activity of the conidiogenone diterpenes. In 2013 five new nitrogen-containing sorbicillinoids, sorbicillamines A–E (**29–33**) (Figure 3) were obtained from the PYG liquid culture of this fungal strain. Despite of their interesting structures no cytotoxic activity (HeLa, BEL7402, HEK-293, and HCT116 cell lines) was detected for these metabolites [27]. Guided by the OSMAC approach, in 2015 the same *Penicillium* species afforded another five new ambuic acid analogues named penicyclones A–E (**34–38**) (Figure 3) , which exhibited antibacterial activities against *Staphylococcus aureus* with MIC values ranging from 0.3 to 1.0 µg/mL [28].

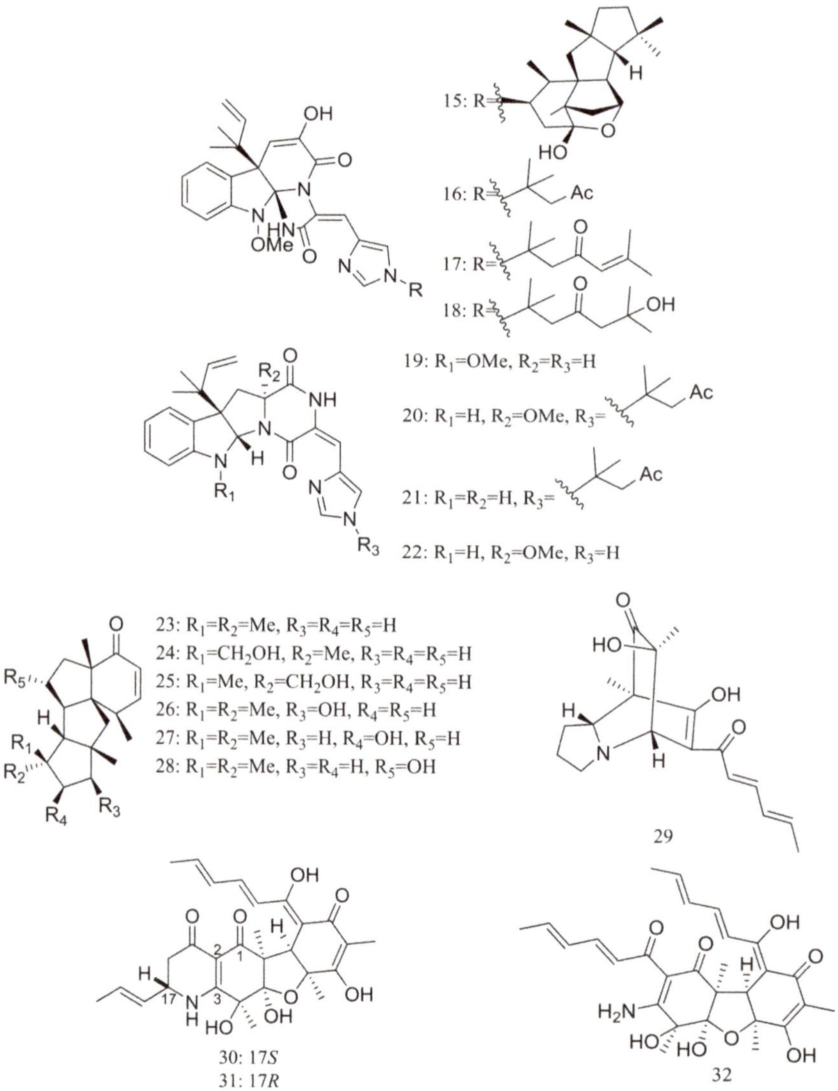

Figure 3. *Cont.*

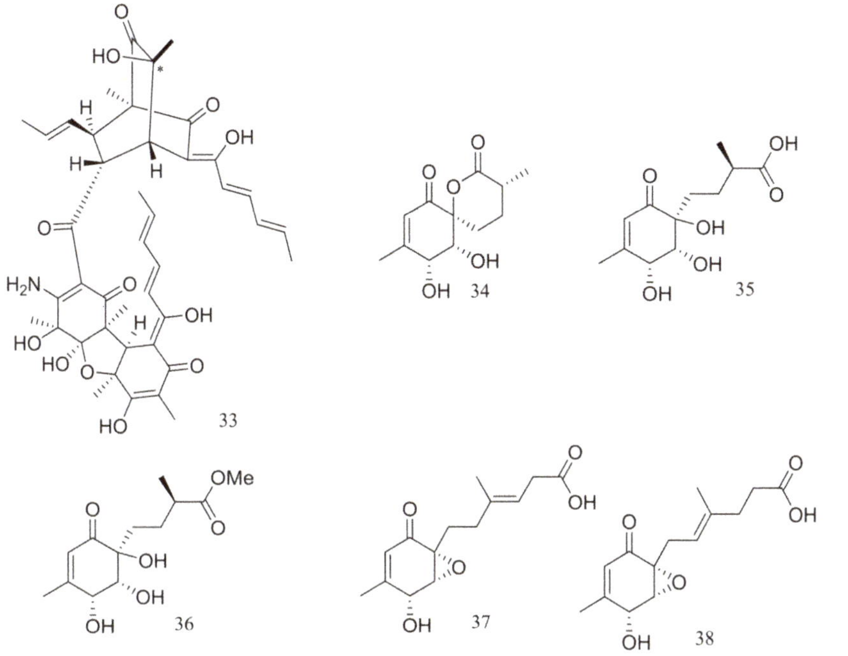

Figure 3. Novel natural products derived from piezophilic fungi (compounds **15–38**). * Absolute configuration is not determined.

Breviones F–H (**39–41**) (Figure 4) were produced by the deep-sea sediment-derived *Penicillium* sp. (MCCC 3A00005) (depth 5115 m, the east Pacific). These three new breviane spiroditerpenoids exhibited cytotoxic activity against HeLa cells with inhibitory rates of 25.2%, 44.9%, and 25.3% at 10 μg/mL, respectively. Effects on HIV-1 inhibition in C8166 cells were tested and an EC_{50} value for compound **39** was 14.7 μM [29]. From the same *Penicillium* strain one new polyoxygenated sterol named sterolic acid (**42**) and three new breviane spiroditerpenoids namend breviones I–K (**43–45**) (Figure 4) were published later. Compound **43** exhibited cytotoxic activity against MCF7 and A549 cells with IC_{50} values of 7.44 and 32.5 μM respectively [30].

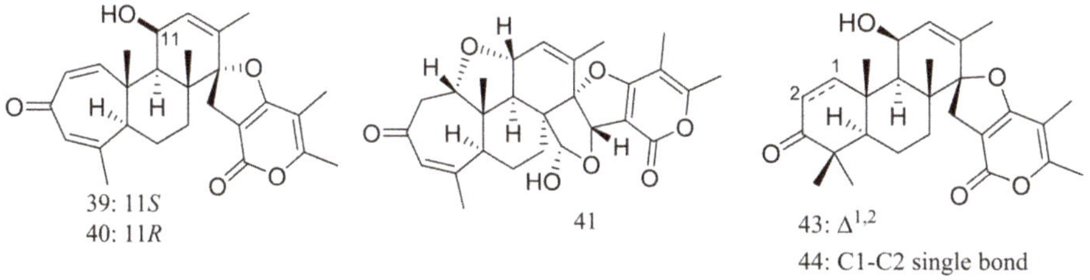

Figure 4. *Cont.*

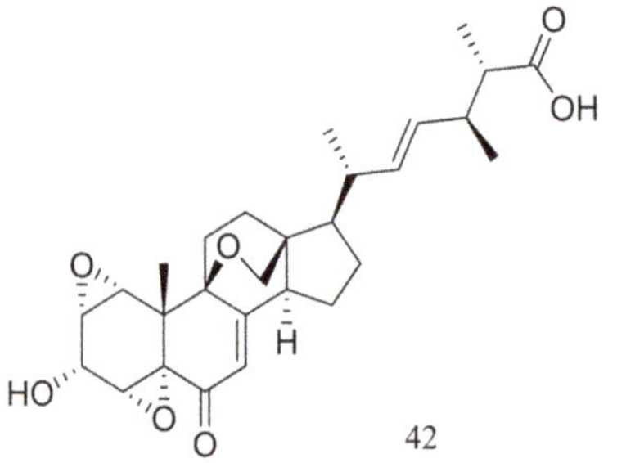

Figure 4. Novel natural products derived from piezophilic fungi (compounds **39–45**).

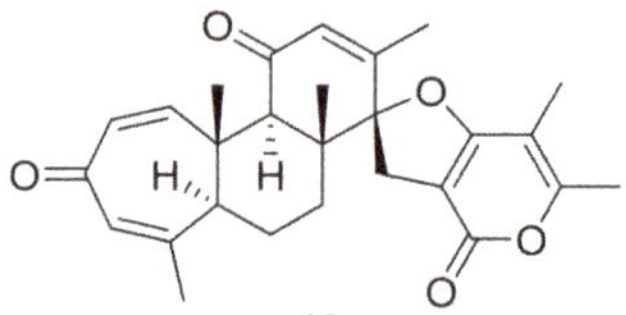

Ascomycotin A (**46**) (Figure 5) was isolated from the deep-sea sediment-derived *Ascomycota* sp. Ind19F07 (depth 3614 m, the Indian Ocean) grown on the rice solid media. No antimicrobial activity (*Acinetobacter baumannii* (ATCC 19606), *Klebsiella pneumoniae* (ATCC 13883), *Escherichia coli* (ATCC 25922), *Staphyloccocus aureus* (ATCC 29213) and *Enterococcus faecalis* (ATCC 29212)) was detected [31].

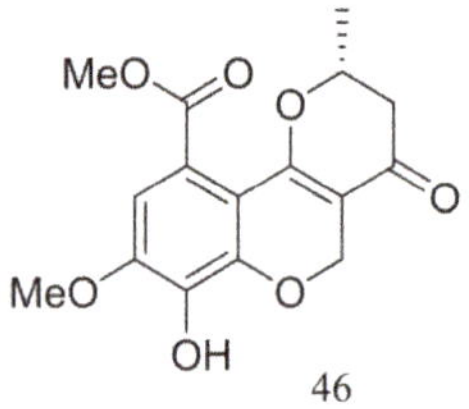

Figure 5. Novel natural products derived from piezophilic fungi (compound **46**).

Cyclopiamides B–J (**47–55**) (Figure 6), nine new cyclopiamide analogues belonging to oxindole alkaloids were produced by the deep-sea-derived fungal strain *Penicillium commune* DFFSCS026 (depth 3563 m, the South China Sea). Toxic activities against brine shrimp of all nine compounds were almost the same with IC_{50} values ranging from 14.1 to 38.5 µg/mL, which suggested that structural modifications at the indole system might not significantly affect their toxic activities. No cytotoxic (HepG-2 and HeLa cell lines) or antiviral (N1H1) activities were detected [32].

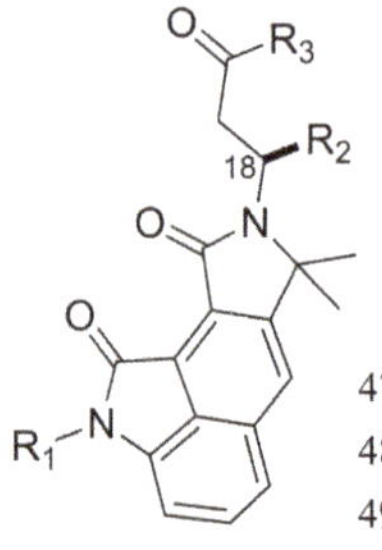

47: 18*, R_1=R_2=Me, R_3=OH
48: 18*, R_1=H, R_2=Me, R_3=OH
49: R_1=H, R_2= =O, R_3=Me

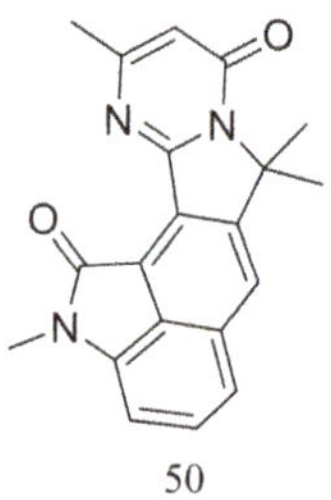

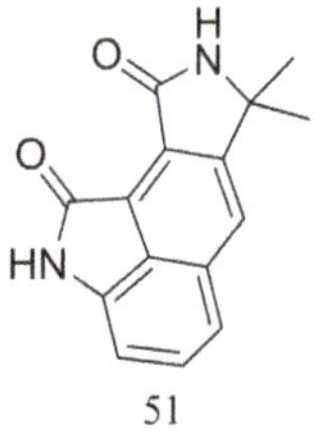

Figure 6. *Cont.*

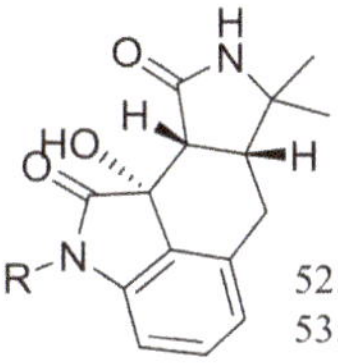
52: R=H
53: R=Me

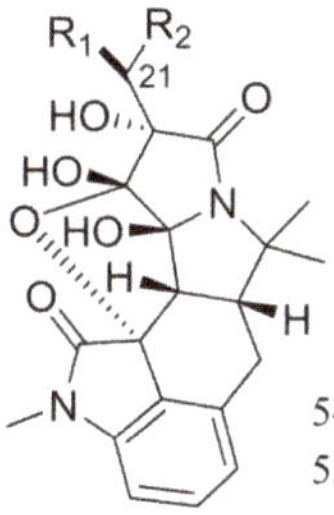
54: 21*S*, R_1=OH, R_2=Me
55: R_1=H, R_2=Ac

Figure 6. Novel natural products derived from piezophilic fungi (compounds **47–55**). * Absolute configuration is not determined.

A new hydroxyphenylacetic acid named westerdijkin A (**56**) (Figure 7) was isolated from a deep-sea sediment-derived fungal strain *Aspergillus westerdijkiae* SCSIO 05233 (depth 4593 m, the South China Sea). Neither antimicrobial (*Escherichia coli*, *Bacillus subtilis*, *Bacillus pumilus*, *Staphylococcus aureus*, and *Candida albicans*), anticancer (K562 and HL-60 cell lines), nor antifouling (*Balanus amphitrite*) activities were detected [33].

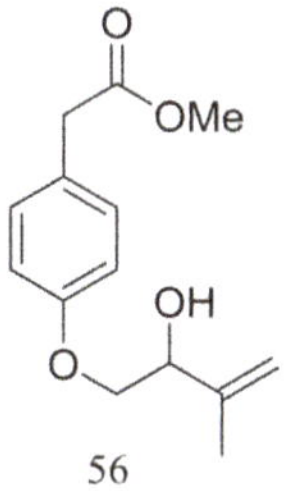
56

Figure 7. Novel natural products derived from piezophilic fungi (compound **56**).

Four new prenylxanthones, emerixanthones A–D (**57–60**) (Figure 8) were isolated from *Emericella* sp. SCSIO 05240 (depth 3258 m, the South China Sea). Compounds **57** and **59** exhibited weak antibacterial activities against six pathogens (*Escherichia coli* (ATCC 29922), *Klebsiella pneumoniae* (ATCC 13883), *Staphylococcus aureus* (ATCC 29213), *Enterococcus faecalis* (ATCC 29212), *Acinetobacter baumannii* (ATCC 19606), and *Aeromonas hydrophila* (ATCC 7966), while **60** displayed mild antifungal activity against six agricultural pathogens (*Fusarium* sp., *Penicillium* sp., *Aspergillus niger*, *Rhizoctonia solani*, *Fusarium oxysporum f.* sp. *niveum*, and *Fusarium oxysporum f.* sp. *cucumeris*). The biosynthetic pathway of these metabolites was proposed [34].

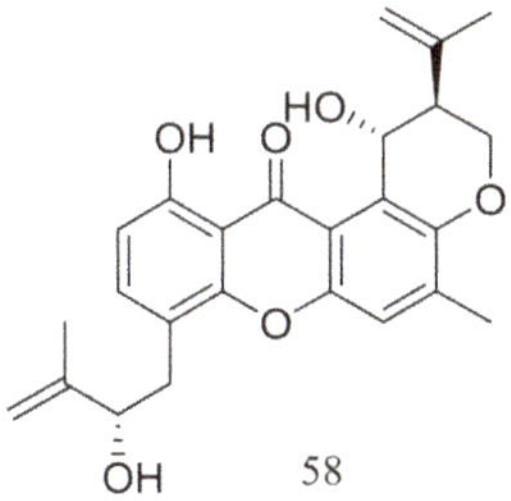
58

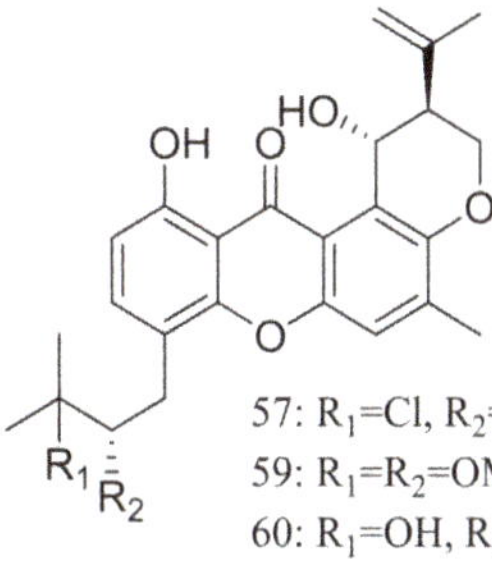
57: R_1=Cl, R_2=OH
59: R_1=R_2=OMe
60: R_1=OH, R_2=OAc

Figure 8. Novel natural products derived from piezophilic fungi (compounds **57–60**).

Engyodontiumones A–J (**61–70**) and 2-methoxyl cordyol C (**71**) (Figure 9) have been described as metabolites of *Engyodontium album* DFFSCS021 taken from a 3739 m deep-sea sediment sample in the South China Sea. Compound **68** exhibited cytotoxic activity against U937 cells (IC$_{50}$ 4.9 µM) and antimicrobial activity against *Escherichia coli* and *Bacillus subtilis* at a concentration of 25 µg/disc [35].

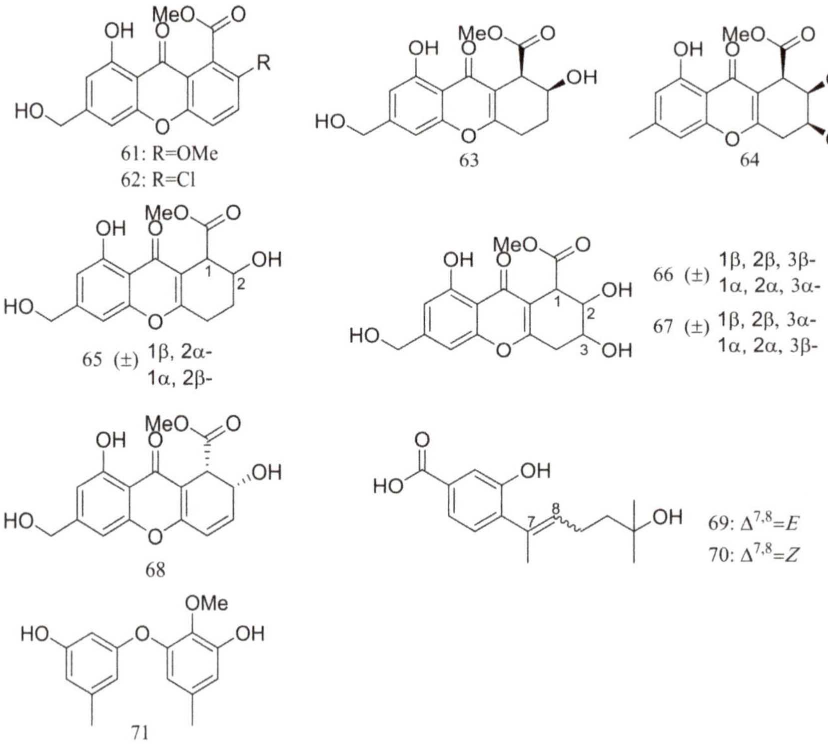

Figure 9. Novel natural products derived from piezophilic fungi (compounds **61–71**).

The solid fermentation of the deep-sea fungus *Aspergillus* sp. SCSIO Ind09F01 (depth 4530 m, the Indian Ocean) yielded a new xanthone named sydoxanthone C (**72**) and a new alkaloid named acremolin B (**73**) (Figure 10). Two compounds exhibited no cytotoxic (Hela, DU145, and U937 cell lines) or COX-2 inhibitory activities [36].

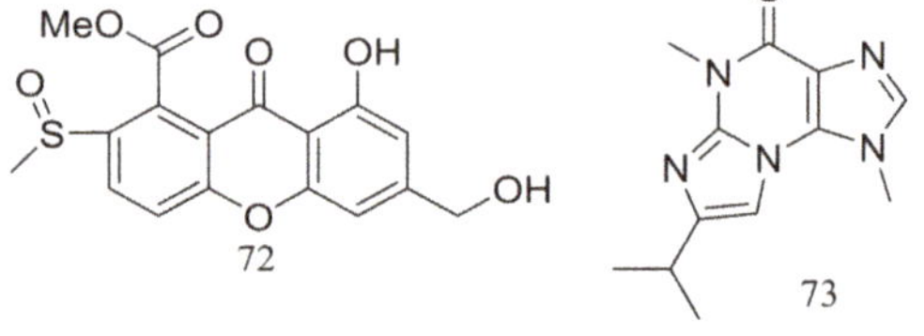

Figure 10. Novel natural products derived from piezophilic fungi (compounds **72–73**).

Dichotocejpins A–C (**74–76**) (Figure 11) are three new diketopiperazines produced by *Dichotomomyces cejpii* FS110 (depth 3941 m). The inhibitory activity of compound **74** (IC_{50} 138 μM) against α-glucosidase was much stronger than that of the positive control acarbose (IC_{50} 463 μM) [37].

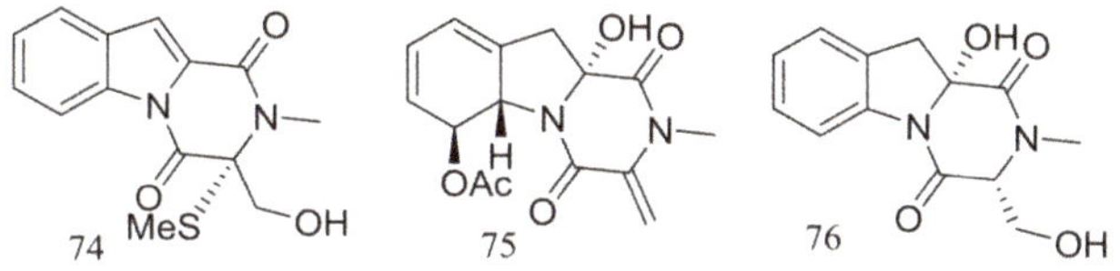

Figure 11. Novel natural products derived from piezophilic fungi (compounds **74–76**).

Acaromycin A (**77**) (Figure 12), a new naphtha-[2,3-b] pyrandione analogue and acaromyester A (**78**) (Figure 12), a new thiazole analogue were isolated from *Acaromyces ingoldii* FS121 (depth 3415 m, the South China Sea). Pronounced cytotoxic activities against four cancer cell lines (MCF-7, NCI-H460, SF-268, and HepG-2) were described for compound **77** with IC_{50} values less than 10 μM [38].

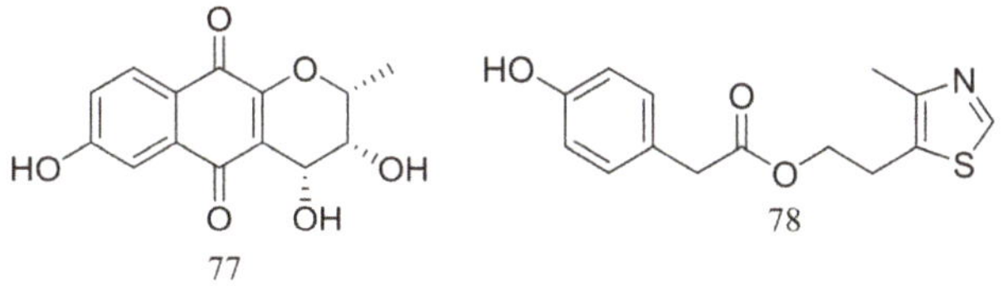

Figure 12. Novel natural products derived from piezophilic fungi (compounds **77–78**).

The trimeric peniphenylanes A–B (**79–80**) and dimeric peniphenylanes C–G (**81–85**) (Figure 13) are seven new 6-methylsaligenin derivatives obtained from *Penicillium fellutanum* HDN14-323 (depth 5752 m, the Indian Ocean). When tested for cytotoxic activity compound **82** proved to be the best active to HeLa cells (IC_{50} 9.3 μM) [39].

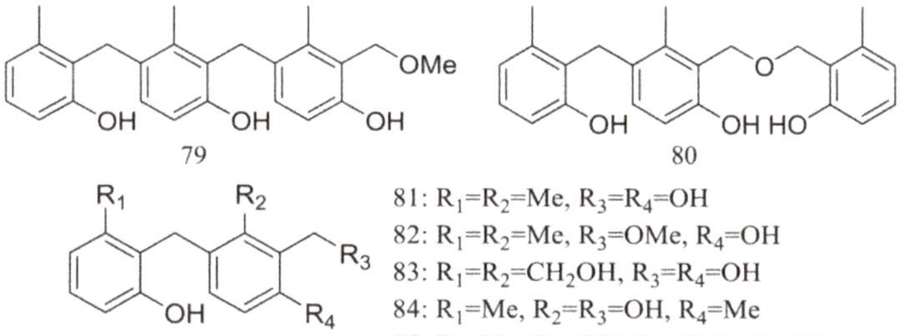

81: R_1=R_2=Me, R_3=R_4=OH
82: R_1=R_2=Me, R_3=OMe, R_4=OH
83: R_1=R_2=CH_2OH, R_3=R_4=OH
84: R_1=Me, R_2=R_3=OH, R_4=Me
85: R_1=Me, R_2=OH, R_3=OMe, R_4=Me

Figure 13. Novel natural products derived from piezophilic fungi (compounds **79–85**).

The deep-sea sediment-derived fungus *Aspergillus versicolor* SCSIO 05879 (depth 3972 m, the Indian Ocean) was found to produce two new oxepine-containing diketopiperazine-type alkaloids named versicoloids A–B (**86–87**) (Figure 14), two new 4-aryl-quinolin-2-one alkaloids (**88–89**) (Figure 14), and four new prenylated xanthones named versicones A–D (**90–93**) (Figure 14). Compounds **86** and **87** displayed the same MIC valued of 1.6 μg/mL against *Colletotrichum acutatum* [40].

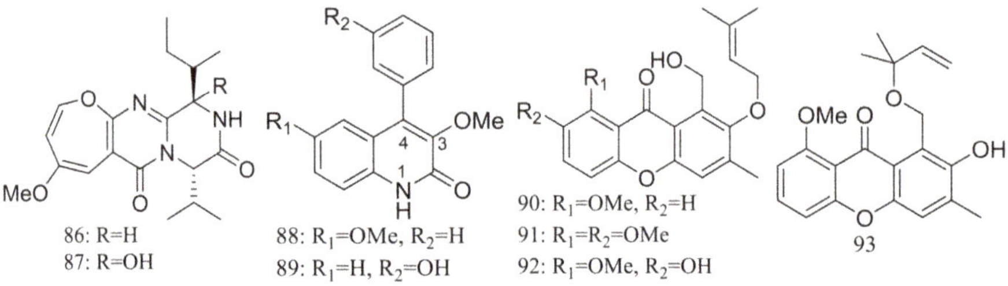

Figure 14. Novel natural products derived from piezophilic fungi (compounds **86–93**).

Aspergilols A–F (**94–99**) (Figure 15) were isolated from fermentations of the deep-sea fungus *Aspergillus versicolor* (A-21-2-7) (depth 3002 m, the South China Sea). Compound **98** significantly activated the Nrf2, which regulated the expression of antioxidant proteins that protect against oxidant damage [41].

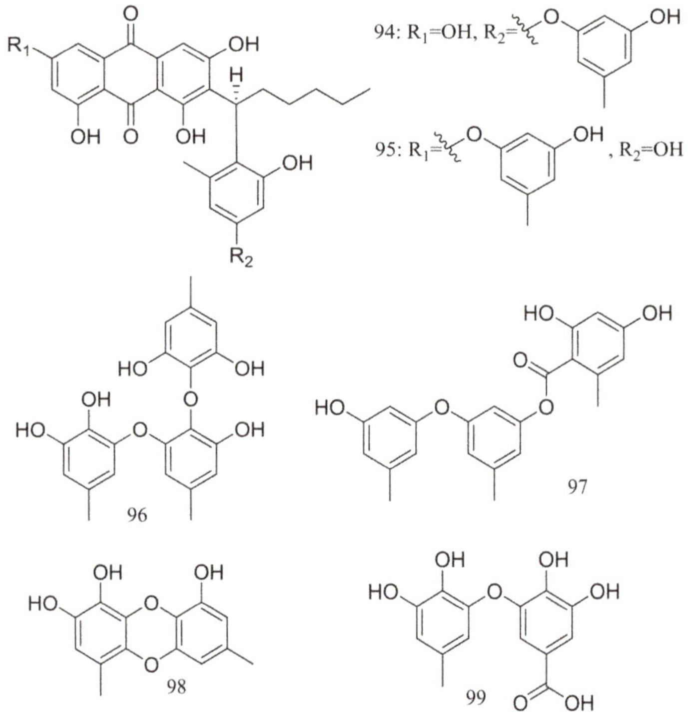

Figure 15. Novel natural products derived from piezophilic fungi (compounds **94–99**).

The clindanones A–B (**100–101**) and cladosporols F–G (**102–103**) (Figure 16) are four new polyketides isolated from the deep-sea fungus *Cladosporium cladosporioides* HDN14-342 (depth 3471 m, the Indian Ocean). Compounds **102–103** showed moderate cytotoxic activity against HeLa, K562, and HCT-116 cell lines with IC$_{50}$ values of 3.9 to 23.0 μM [42].

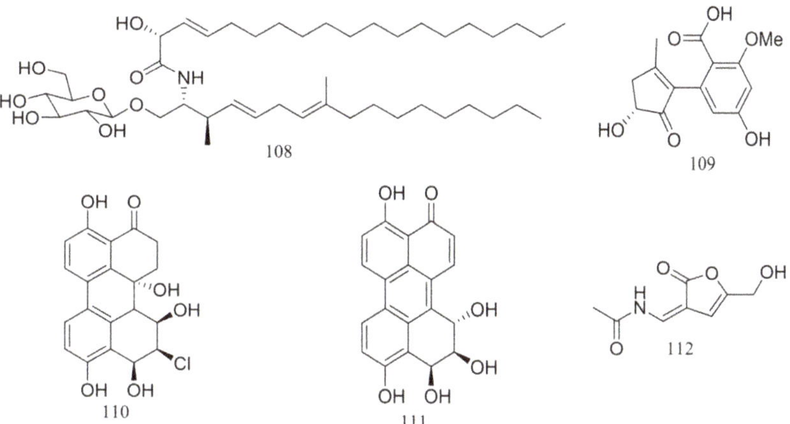

Figure 16. Novel natural products derived from piezophilic fungi (compounds **100–103**).

The fungal strain *Penicillium brevicompactum* DFFSCS025 (depth 3928 m, the South China Sea) produced two new brevianamides, brevianamids X–Y (**104–105**) and two new mycochromenic acid derivatives (**106–107**) (Figure 17). Compound **106** showed strong antilarval activity against *Bugula neritina* with an EC$_{50}$ value of 13.7 µM [43].

Figure 17. Novel natural products derived from piezophilic fungi (compounds **104–107**).

In exploring for new BRD4 inhibitors, five new compounds including one new cerebroside (**108**) (Figure 18), one new alternaric acid (**109**) (Figure 18), two new perylenequinones (**110–111**) (Figure 18), and 2-(*N*-vinylacetamide)-4-hydroxymethyl-3-ene-butyrolactone (**112**) (Figure 18) were isolated from fermentations of *Alternaria* sp. NH-F6 (depth 3927 m, the South China Sea). Compound **111** was a potent inhibitor with an inhibition rate of 88.1% at 10 µM, while **110** had a moderate inhibition at rate of 57.7% at the same concentration [44].

Figure 18. Novel natural products derived from piezophilic fungi (compounds **108–112**).

Engyodontiumin A (**113**) (Figure 19) was produced by the deep-sea-derived fungus *Engyodontium album* (depth 3542 m, the Atlantic Ocean). This novel benzoic acid derivative displayed moderate antibacterial activity against *Aspergillus niger*, MRSA, *Vibrio vulnificus*, *Vibrio rotiferianus*, and *Vibrio campbellii*. The experimental data on the antimicrobial activity were not provided in the original article [45].

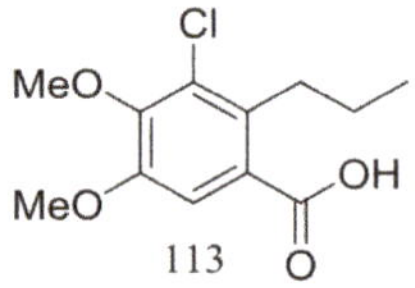

Figure 19. Novel natural products derived from piezophilic fungi (compound **113**).

In exploration for novel bioactive marine natural products, four new isobenzofuanones named leptosphaerins JM (**114–117**) and two new isochromenones named clearanols I–J (**118–119**) (Figure 20) were isolated from *Leptosphaeria* sp. SCSIO 41005 (depth 3614 m, the Indian Ocean). When evaluated for biological activity, no cytotoxicity (K562, MCF-7, and SGC7901 cell lines) or antiviral activity (H3N2, EV71, and HIV viruses) was detected [46].

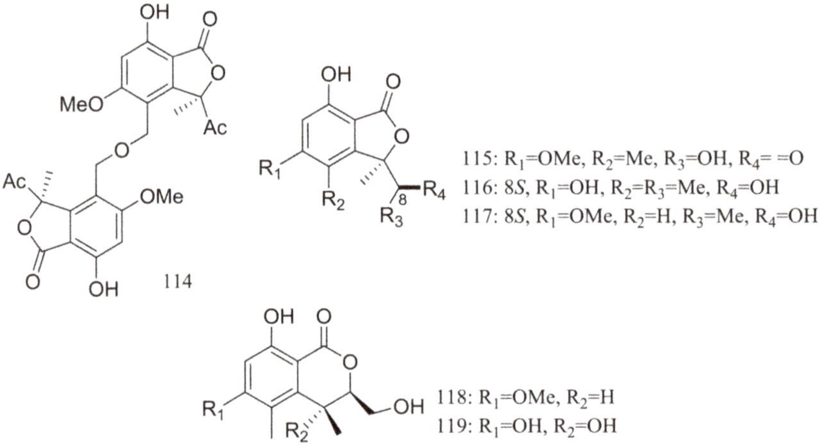

Figure 20. Novel natural products derived from piezophilic fungi (compounds **114–119**).

A mixed culture of the deep-sea-derived fungus *Talaromyces aculeatus* (depth 3386 m, the Indian Ocean) and the mangrove-derived fungus *Penicillium variabile* (Fujian Province of China) afforded four new polyketides, penitalarins A–C (**120–122**) and nafuredin B (**123**) (Figure 21). None of these compounds was produced by either of the two fungi when cultured alone under the same conditions. Compound **123** inhibited a panel of cancer cell lines (HeLa, K562, HCT-116, HL-60, A549, and MCF-7) with IC_{50} values ranging from 1.2 to 9.8 μM (doxorubicin as positive control IC_{50} 0.2 to 0.8 μM) [47].

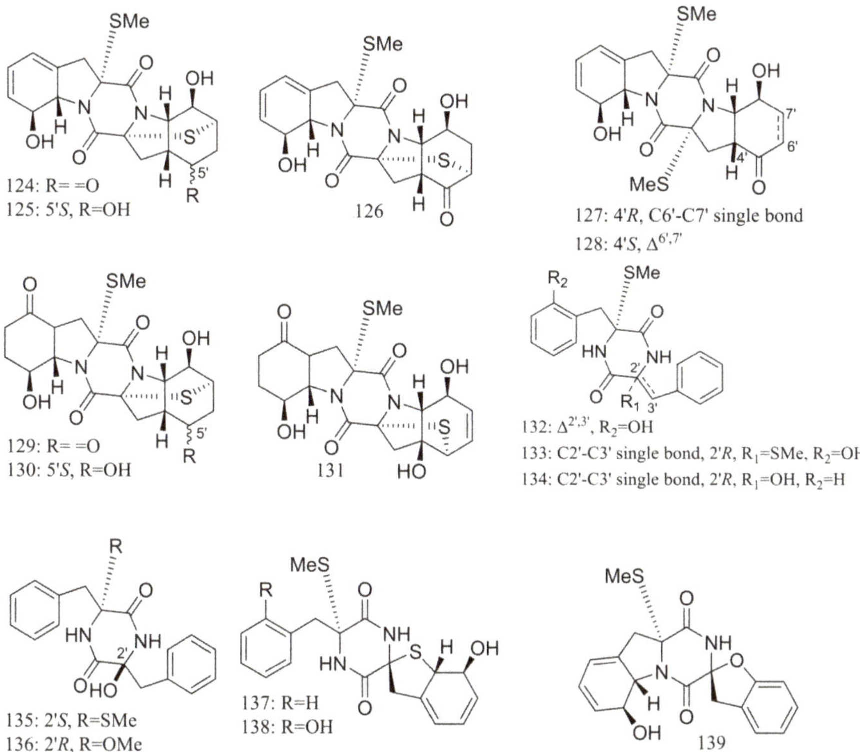

Figure 21. Novel natural products derived from piezophilic fungi (compounds **120–123**).

Nineteen new thiodiketopiperazine-type alkaloids named eutypellazines A–S (**124–142**) (Figure 22) [48,49] were produced by the marine-derived fungus *Eutypella* sp. MCCC 3A00281 (depth 5610 m, the South Atlantic Ocean). Inhibitory effects on HIV-1 replication in pNL4.3Env-.Luc co-transfected 293T cells were tested and IC_{50} values for compounds **124–135** ranged from 3.2 to 18.2 μM (EFV as the positive control IC_{50} 0.1 μM). In addition, compound **133** could reactivate latent HIV-1 in J-Lat A2 cells in a dose-dependent manner. When tested for antimicrobial activity compounds **139–142** were active to *Staphylococcus* aureus ATCC 25923 and vancomycin-resistant enterococci with MIC values of 32/32, 16/16, 32/32, and 16/32 μM respectively.

Figure 22. *Cont.*

Figure 22. Novel natural products derived from piezophilic fungi (compounds **124–142**).

The cultured broth of the sea cucumber-derived fungus *Penicillium coralligerum* YK-247 (depth 3064 m, São Paulo Plateau, off Brazil) potently inhibited the growth of *Saprolegnia parasitica*. Further chromatographic fractionation of the cultured broth led to the isolation of cladomarine (**143**) (Figure 23) which showed selective antimicrobial activity against *Saprolegnia parasitica* and *Pythium* sp. sakaril at a concentration of 10 µg/disc [50].

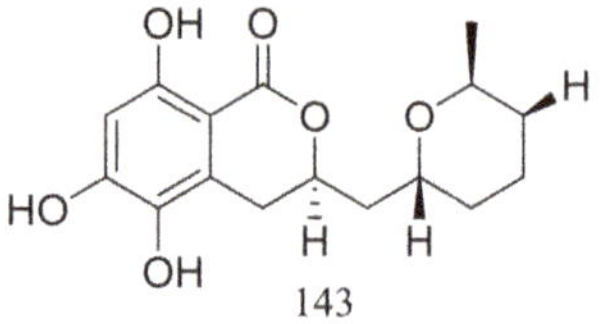

Figure 23. Novel natural products derived from piezophilic fungi (compound **143**).

3. Psychrophilic Fungi

Psychrophilin D (**144**) (Figure 24), a new cyclic nitropeptide was isolated from *Penicillium algidum* derived from a soil sample in Greenland. This compound exhibited moderate cytotoxic activity against P388 murine leukaemia cells with an ID_{50} value of 10.1 µg/mL. When evaluated for antimicrobial, antiviral, anticancer and antiplasmodial activities compound **144** proved to be inactive [51].

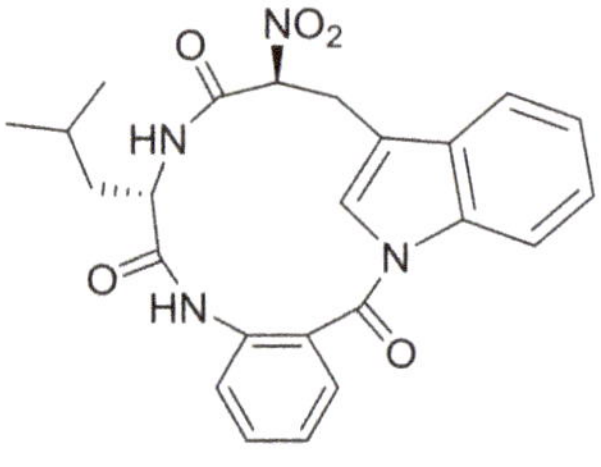

Figure 24. Novel natural products derived from psychrophilic fungi (compound **144**).

In 2005 Oh et al. discovered libertellenones A–D (**145–148**) (Figure 25) when co-cultured a marine-derived fungus with a unicellular marine bacterium. Libertellenone D (**148**) demonstrated potent cytotoxicity (IC_{50} 0.76 µM) against HCT-116 cell line, whereas the other libertellenones exhibited weaker activities (IC_{50} 15, 15 and 53 µM respectively) [52]. In 2014 libertellenones G (**149**) and H (**150**) (Figure 25) together with **145** and **147** were isolated from *Eutypella* sp. D-1, which was derived from a soil sample collected on London Island of Kongsfjorden of Ny-Ålesund District (altitude of 100 m), Arctic. Compound **149** showed moderate antibacterial activity against *Escherichia coli*, *Bacillus subtilis*, and *Staphylococcus aureus*. Compound **150** showed slight cytotoxicity against several cancer cell lines (MCF-7, H460, U251, SW-1990, Hela, Huh-7, and SG7901) with IC_{50} values between 3.31 and 44.1 µM.

According to the structure-bioactivity relations, the cyclopropane ring in **148** and **150** appears to be an important structural feature associated with their biological activity [53]. Later Chu's group found that **150** biosynthesis was significantly elevated (16.4 folds) with ethanol treatment, and further study showed that the gene transcription levels of 3-hydroxy-3-methyl glutaric acyl coenzyme A reductase and geranylgeranyl diphosphate synthase were up-regulated by ethanol stimulation [54]. Several new compounds including cytochalasins Z$_{24}$, Z$_{25}$, Z$_{26}$ (**151–153**) (Figure 25) [55], eutypenoids A–C (**154–156**) (Figure 25) [56], and *eut*-Guaiane sesquiterpene (**157**) (Figure 25) [57] have been described from the same fungal strain since 2014. Compound **151** exhibited a moderate cytotoxicity against MCF-7 cells with an IC$_{50}$ value of 9.33 µM. Compound **155** was able to suppress the proliferation of BALB/c mice splenocytes under ConA induction. Antibacterial activity (*Escherichia coli*, *Bacillus subtilis*, and *Staphylococcus aureus*) of compound **157** was comparable to that of ampicillin but cytotoxic activity against SGC7901 cells was very weak (IC$_{50}$ 39.8 µM).

Figure 25. *Cont.*

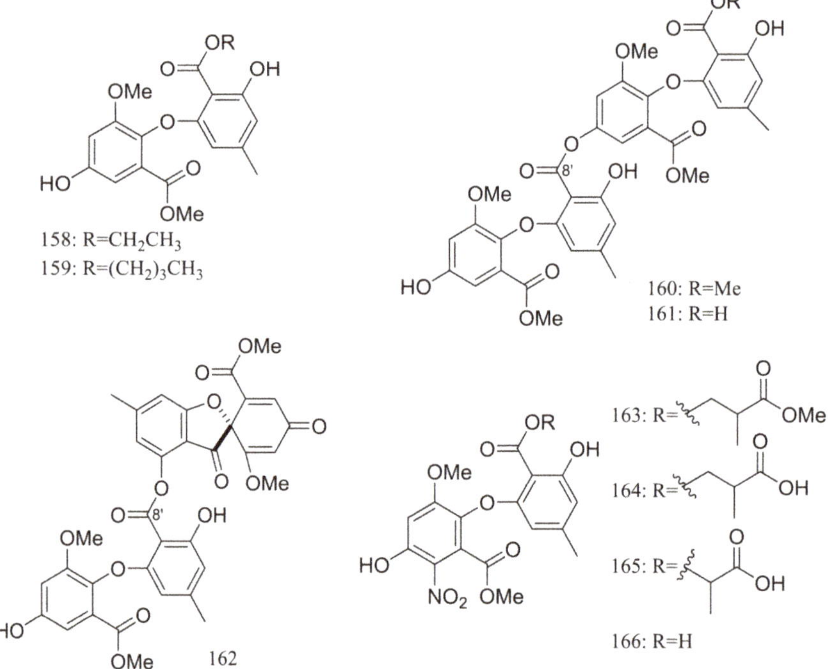

Figure 25. Novel natural products derived from psychrophilic fungi (compounds **145–157**).

In search for new antifungal and antibacterial natural products, five asterric acid derivatives named ethyl asterrate (**158**), n-butyl asterrate (**159**) and geomycins A–C (**160–162**) (Figure 26) were isolated from an Antarctic *Geomyces* species. Compound **161** showed significant antifungal activity against *A. fumigatus* (ATCC 10894) with IC_{50}/MIC values of 0.86/29.5 µM (the positive control fluconazole IC_{50}/MIC 7.35/163.4 µM). Compound **162** exhibited moderate antimicrobial activity against both Gram-positive and Gram-negative bacteria with IC_{50} values ranging from 12.9 to 36.2 µM [58]. In 2015 four nitroasterric acid derivatives named pseudogymnoascins A–C (**163–165**) and 3-nitroasterric acid (**166**) (Figure 26) were described as metabolites of a sponge-associated fungus *Pseudogymnoascus* sp. F09-T18-1, which was collected from the King George Island of Antarctic. No antimicrobial activity was observed at MIC > 64 µg/mL. Compared with compounds **161** and **162**, the lack of antimicrobial activities of compounds **163–166** suggested the activity lied in the size of substituent at C-8′ and/or the presence of the nitro group in the molecule [59].

Figure 26. Novel natural products derived from psychrophilic fungi (compounds **158–166**).

Several piperazine-type compounds, chetracins B–D (**167–169**) and oidioperazines A–D (**170–173**) (Figure 27) were produced by the soil-derived Antarctic fungus *Oidiodendron truncatum* GW3-13 which was obtained near the Great Wall station (Chinese Antarctic station). When tested for cytotoxic activities against a panel of cancer cell lines (HCT-8, Bel-7402, BGC-823, A549, and A2780) compound **167** proved to be the most active (IC$_{50}$ 0.003 to 0.028 µM), whereas **168** and **169** were less active (IC$_{50}$ 0.14 to 1.83 µM) [60].

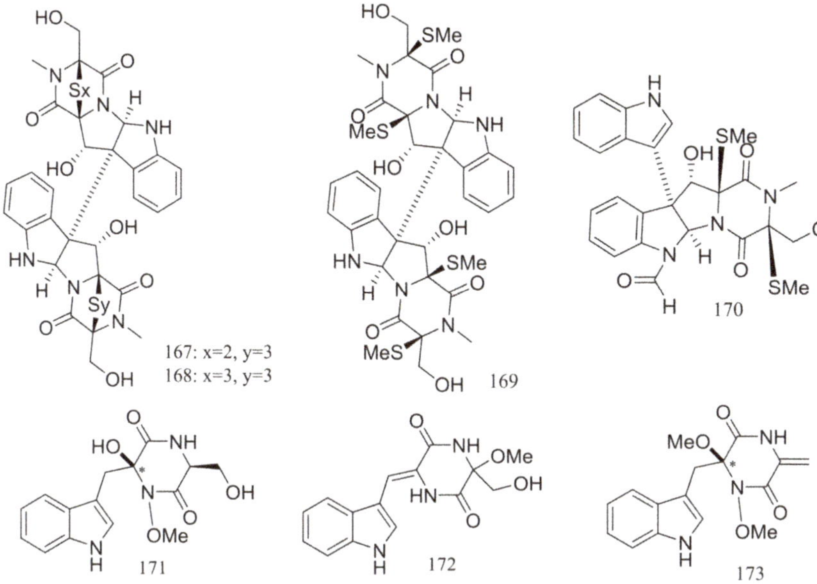

Figure 27. Novel natural products derived from psychrophilic fungi (compounds **167–173**). * Absolute configuration is not determined.

Two highly oxygenated polyketides, penilactones A and B (**174** and **175**) (Figure 28) featuring a new carbon skeleton formed from two 3,5-dimethyl-2,4-diol-acetophenone units and a γ-butyrolactone moiety were produced by the Antarctic marine-derived fungus *Penicillium crustosum* PRB-2. A plausible biogenetic pathway was proposed in the original article. Effects on NF-κB inhibition (in transient transfection and reporter gene expression assay) in RAW264.7 cells were tested and compound **174** showed very weak activity with a rate of 40% at 10 mM (the positive control PDTC inhibitory rate 85% at 0.1 mM) [61].

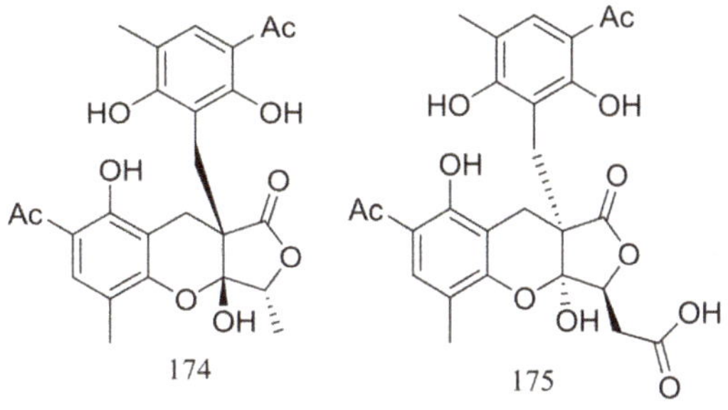

Figure 28. Novel natural products derived from psychrophilic fungi (compounds **174–175**).

Several interesting eremophilane-type sesquiterpenes with high structural diversity have been described for *Penicillium* sp. PR19N-1 derived from a sludge sample in Prydz Bay (−1000 m), Antarctica. In 2013 four new chloro-eremophilane sesquiterpenes (**176–179**) (Figure 29) were isolated from this fungal strain and the plausible metabolic network was proposed. Compound **176** displayed modest cytotoxic activity against HL-60 and A549 cell lines with IC_{50} values of 11.8 and 12.2 µM respectively, whereas the other compounds exhibited no activities. [62]. Soon later another five new eremophilane-type sesquiterpenes (**180–184**) and a rare lactam-type eremophilane (**185**) (Figure 29) were isolated from the same *Penicillium* strain. When tested for cytotoxic activities against HL-60 and A-549 cells only **180** and **184** proved to be active and compound **184** displayed strong cytotoxic activity against A-549 cells with an IC_{50} value of 5.2 µM [63].

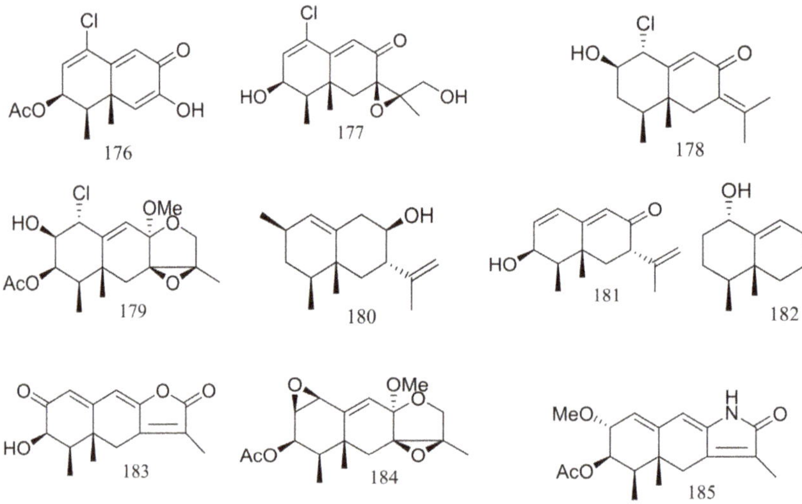

Figure 29. Novel natural products derived from psychrophilic fungi (compounds **176–185**).

Two different Lindgomycetaceae strains KF970 and LF327 obtained from different marine habitats (Antarctic and the Kiel Fjord, Baltic Sea) both produced lindgomycin (**186**) (Figure 30), an unusual polyketide with a unique 5-benzylpyrrolidine-2,4-dione unit at the tetramic acid substructure. Antibiotic activity (*Bacillus subtilis*, *Staphylococcus aureus*, and methicillin-resistant *Staphylococcus aureus*) of compound **186** were two times less than that of chloramphenicol (the positive control) [64].

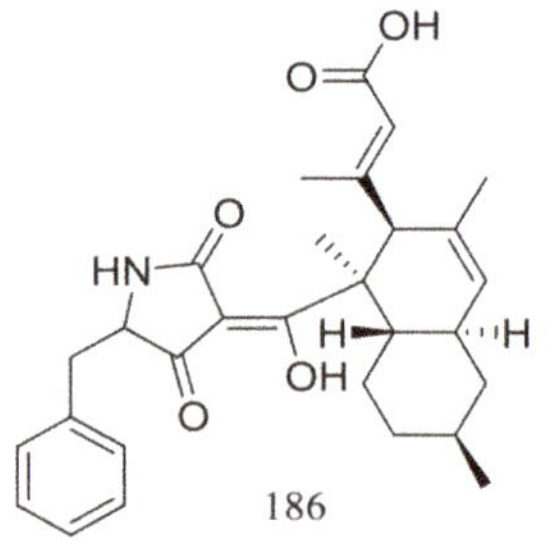

Figure 30. Novel natural products derived from psychrophilic fungi (compound **186**).

The psychrotolerant fungus *Penicillium* sp. SCSIO 05705 collected nearby the Great Wall station (Chinese Antarctic station) afforded three new indolyl diketopiperazine derivatives, penillines A–B (**187–188**) and isopenilline A (**189**) (Figure 31). In the general bioactivity profiling programs including antiviral, cytotoxic, antibacterial and antituberculosis evaluation, all compounds were found inactive [65].

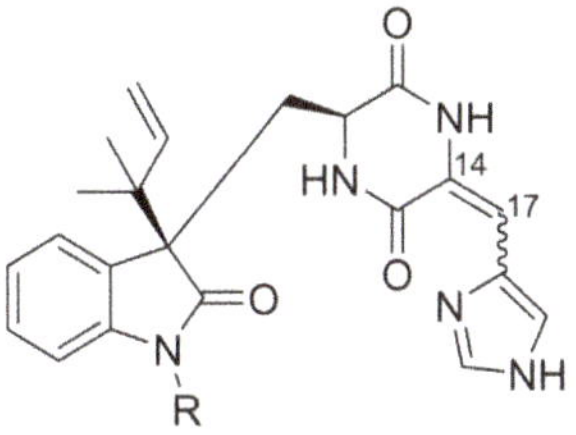

187: $\Delta^{14,17}$=*E*, R=OH,
188: $\Delta^{14,17}$=*E*, R=H,
189: $\Delta^{14,17}$=*Z*, R=OH,

Figure 31. Novel natural products derived from psychrophilic fungi (compounds **187–189**).

Chrodrimanins I (**190**) and J (**191**) (Figure 32), two new meroterpenoids were isolated from the moss-derived *Penicillium funiculosum* GWT2-24, collected at the China Great Wall Station in Antarctica. Distinguished from the reported chrodrimanins, compounds **190** and **191** possessed a unique cyclohexanone instead of a δ-lactone ring. Neither antiviral activity (H1N1) nor cytotoxic activity (K562, HL60, HeLa, and A549 cell lines) was detected [66]. From the same fungal strain, six new pyridine alkaloids named penipyridones A–F (**192–197**) (Figure 32) were published later. When screened for lowering of oleic acid elicited lipid accumulation in HepG2 hepatocytes, compounds **192**, **193** and **196** remarkably reduced intracellular lipid accumulation as well as the total cholesterol and triglyceride quantification at 10 μM [67].

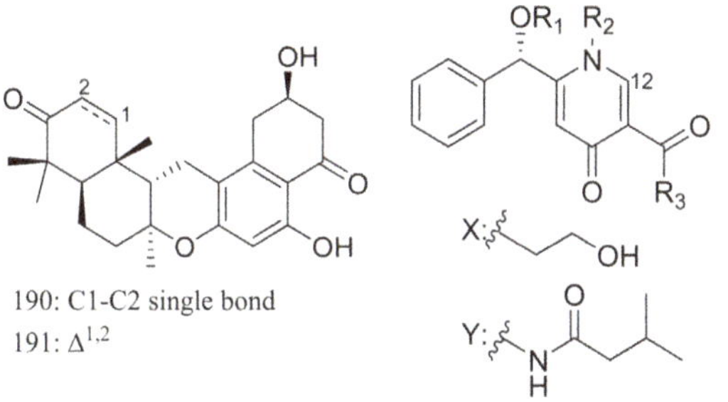

190: C1-C2 single bond
191: $\Delta^{1,2}$

192: R_1=R_2=H, R_3=NH_2
193: R_1=R_2=H, R_3=OMe
194: R_1=R_2=H, R_3=Y
195: R_1=H, R_2=X, R_3=NH_2
196: R_1=Ac, R_2=X, R_3=NH_2
197: R_1=H, R_2=X, R_3=Y

Figure 32. Novel natural products derived from psychrophilic fungi (compounds **190–197**).

Exopisiod B (**198**) and farylhydrazone C (**199**) (Figure 33) were produced by a soil-derived fungus *Penicillium* sp. HDN14-431 collected from mesolittoral zone in Antarctic. Both compounds exhibited no cytotoxicity (K562, A549, HCT116, and HeLa cell lines) at IC_{50} > 10 μM, but compound **199** was slightly against *Proteusbacillus vulgaris* with an MIC value of 22.5 μM [68].

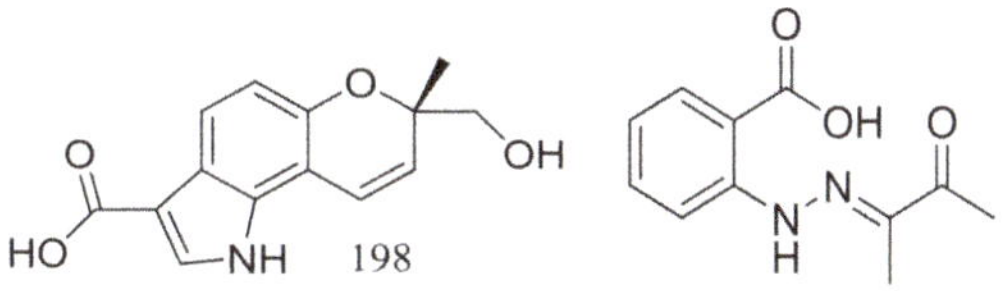

199

Figure 33. Novel natural products derived from psychrophilic fungi (compounds **198–199**).

The soil-derived fungus *Aspergillus ochraceopetaliformis* SCSIO 05702 collected near the Great Wall station (Chinese Antarctic station) yielded five new highly oxygenated α-pyrone merosesquiterpenoids named ochraceopones A–E (**200–204**), the known asteltoxin (**205**), and a new double bond isomer of asteltoxin named isoasteltoxin (**206**) (Figure 34). Compounds **200–203** possessed a linear tetracyclic carbon skeleton, which was distinguished from the reported angular tetracycle structure. Compounds **200**, **205**, and **206** displayed antiviral activity against the H1N1 and H3N2 influenza viruses with IC_{50} values of >20.0/12.2, 0.54/0.84, and 0.23/0.66 μM, respectively (the positive control tamiflu IC_{50} 16.9/18.5 nM). In addition, the selectivity indexes (SI = CC_{50}/IC_{50}) of anti-H1N1 activity of **205** (SI = 0.44) and **206** (SI = 2.35) suggested that the geometry of the Δ^{11} double bond in the polyene chain might accelerate the anti-H1N1 activity and selectivity index [69].

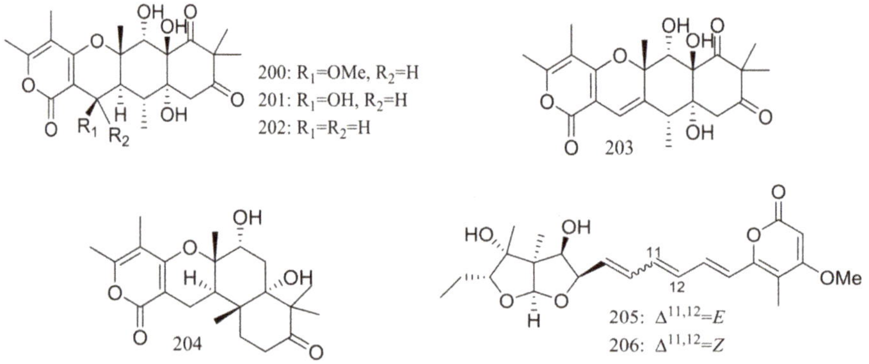

Figure 34. Novel natural products derived from psychrophilic fungi (compounds **200–206**).

A furanone derivative, butanolide A (**207**) and a sesquiterpene, guignarderemophilane F (**208**) (Figure 35) were produced by the Antarctic seabed sediment-derived fungus *Penicillium* sp. S-1-18 via the bioassay guidance. Compound **207** could moderately inhibited protein tyrosine phosphatase 1B (PTP1B) with an IC_{50} value of 27.4 μM [70].

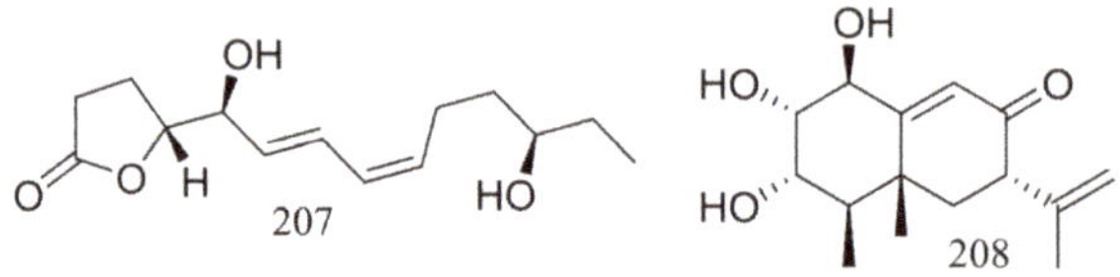

Figure 35. Novel natural products derived from psychrophilic fungi (compounds **207–208**).

Penicillium granulatum MCCC 3A00475 obtained from the Prydz Bay of Antarctica yielded an unusual spirocyclic diterpene named spirograterpene A (**209**) (Figure 36). Antiallergic effect was tested in immunoglobulin E-mediated rat mast RBL-2H3 cells and compound **209** was just little weaker active than loratadine at 20 μg/mL [71].

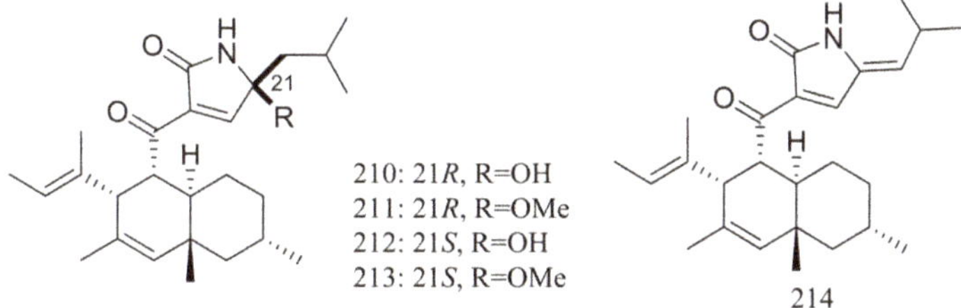

Figure 36. Novel natural products derived from psychrophilic fungi (compound **209**).

4. Thermophilic Fungi

Five new polyketides (**210–214**) (Figure 37) were produced by *Myceliophthora thermophila* obtained from the soil of fumaroles in Taiwan. Compounds **210–212** showed cytotoxic activity against A549, Hep3B, MCF-7 and HepG2 cell lines with IC$_{50}$ values ranging from 0.25 to 1.30 µg/mL [72].

210: 21*R*, R=OH
211: 21*R*, R=OMe
212: 21*S*, R=OH
213: 21*S*, R=OMe

214

Figure 37. Novel natural products derived from thermophilic fungi (compounds **210–214**).

The EtOAc extract of the mass mycelium and PDA media of *Malbranchea sulfurea* which was obtained from the soil of fumaroles in Sihchong River Hot Spring Zone, displayed strong cytotoxicity against several cancer cell lines. Further bioassay-guided fractionation and chromatographic separation of the extract led to the isolation of six photosensitive polyketides named malbranpyrroles A–F (**215–220**) (Figure 38). Cytotoxic activities against PANC-1, HepG2 and MCF-7 cancer cell lines were tested and IC$_{50}$ values for compounds **217–220** ranged from 3 to 11 µM. Flow cytometric measurement for cell cycle analysis showed that when treated by the malbranpyrroles the percentage of MCF-7 and HepG2 cells in G0/G1 phase was slightly increased, and the results suggested that these cytotoxic compounds could arrest the two cancer cell lines at G0 phase via inhibiting some cellular signaling pathways. According to the structure-bioactivity relations, the chlorine atom might be the pharmacophore for cytotoxicity [73].

215

216

217

Figure 38. *Cont.*

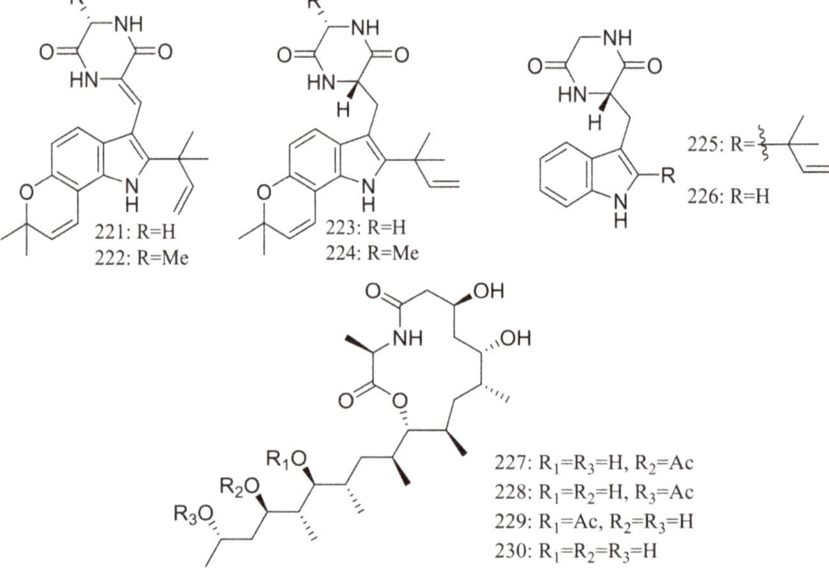

Figure 38. Novel natural products derived from thermophilic fungi (compounds **215–220**).

From two fungal strains *Talaromyces thermophilus* YM1-3 and YM3-4, both collected from Tengchong hot springs, six new indole alkaloids including talathermophilins A–B (**221–222**) (Figure 39) [74], two analogues of notoamide E (**223** and **224**), one analogue of preechinulin (**225**), and a natural occurring cyclo (glycyltryptophyl) (**226**) (Figure 39) [75], as well as a novel class of PKS-NRPS hybrid molecules named thermolides A–F (**226–232**) (Figure 39) [76] have been described. Compounds **221** and **222** exhibited moderate nematicidal activities against the worms of the free-living nematode *Panagrellus redivivus* with rates of 38% and 44% at 400 µg/mL for 72 h respectively. Compounds **227–232** possessed a 13-membered lactam-bearing macrolactone but only **227** and **228** displayed potent nematicidal activities against three notorious nematodes (*Meloidogyne incognita*, *Bursaphelenches siylopilus*, and *Panagrellus redivivus*) with LC$_{50}$ values between 0.5 and 1 µg/mL. No information on the biological activities of compounds **223–226** was given.

Figure 39. *Cont.*

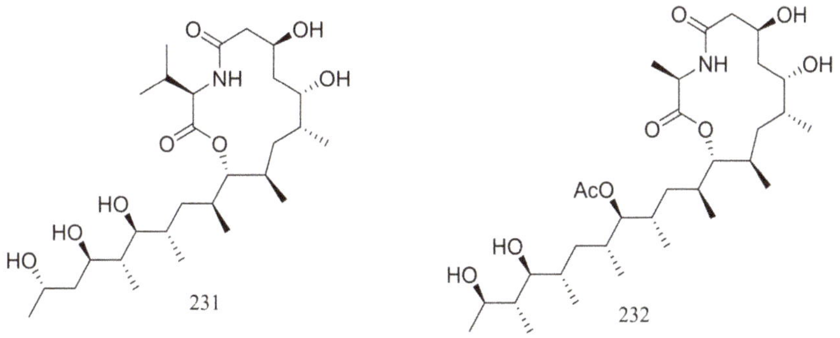

Figure 39. Novel natural products derived from thermophilic fungi (compounds **221–232**).

Clavatustides A–B (**233–234**) (Figure 40) containing an unusual anthranilic acid dimer and a D-phenyllactic acid residues were produced by *Aspergillus clavatus* C2WU isolated from the crab *Xenograpsus testudinatus*, which lived at extreme, toxic habitat around the sulphur-rich hydrothermal vents in Taiwan Kueishantao. The two novel cyclodepsipeptides significantly suppressed the proliferation of HepG2 cells in a dose-dependent manner, and cell cycle analysis suggested that **233** and **234** could induce G1 arrest and inhibit G1/S phase transition [77].

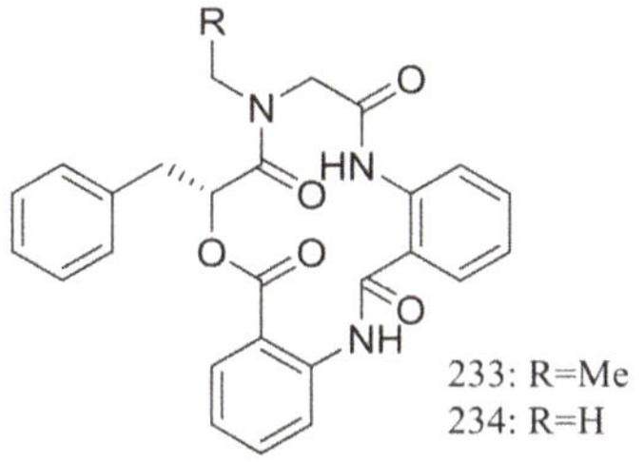

Figure 40. Novel natural products derived from thermophilic fungi (compounds **233–234**).

Nine new C_9 polyketides named aspiketolactonol (**235**), aspilactonols A–F (**236–241**), aspyronol (**242**) and epiaspinonediol (**243**) (Figure 41) have been described as metabolites of *Aspergillus* sp. 16-02-1, which was collected at a Lau Basin hydrothermal vent (depth 2255 m, 114 °C) in the Southwest Pacific. Compounds **240** and **241** were obtained as a mixture in a diastereomeric ratio of 1:1. The possible biosynthetic pathways for all compounds were proposed and discussed. The cytotoxic activities (IC_{50} value) against HL-60 cells of compounds **242** and **243** were 241.2 and 192.9 μM respectively. For compounds **235–241** very weak cytotoxic activities (K562, HL-60, HeLa, or BGC-823) were observed with inhibitory rates less than 20% at 100 μg/mL [78].

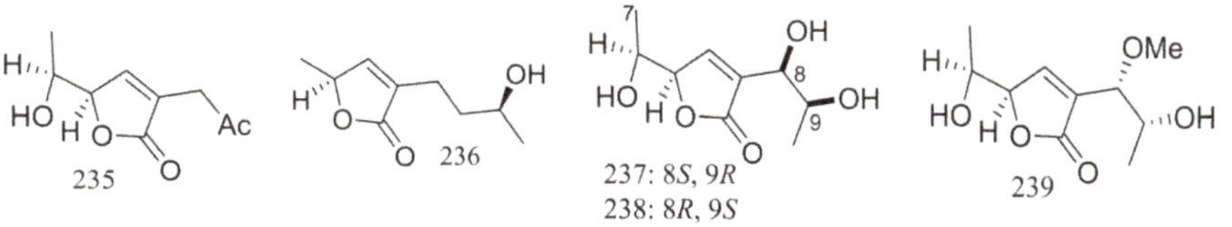

Figure 41. *Cont.*

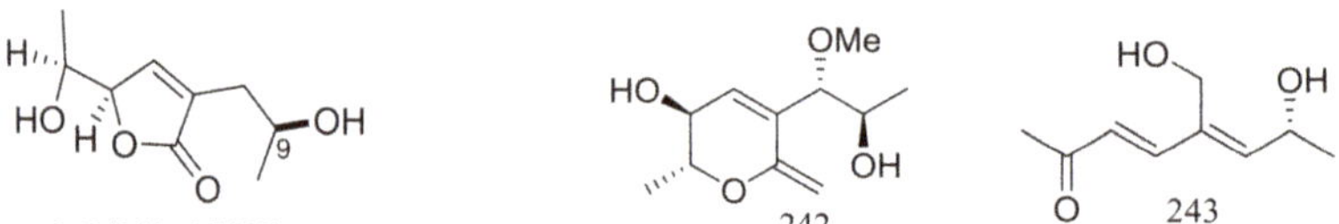

Figure 41. Novel natural products derived from thermophilic fungi (compounds **235–243**).

A hydrothermal vent fungus *Penicillium* sp. Y-50-10 collected from the sulfur rich sediment (Kueishantao, Taiwan) yielded methyl isoverrucosidinol (**244**) (Figure 42). This new verrucosidin derivative displayed weak antibiotic activity against *Bacillus subtilis* with an MIC value of 32 μg/mL [79].

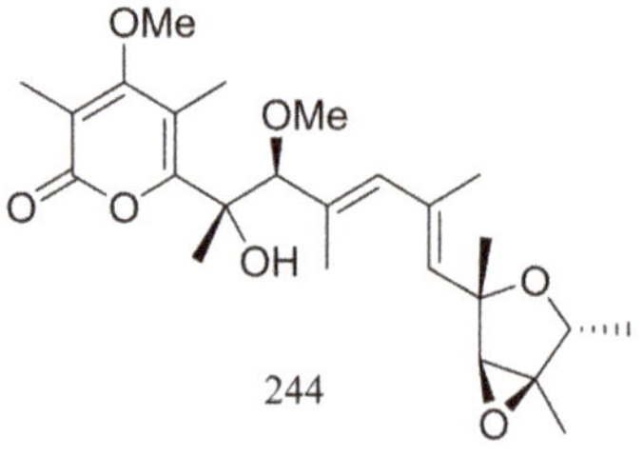

Figure 42. Novel natural products derived from thermophilic fungi (compound **244**).

The soil-derived thermophilic fungal strain *Aspergillus terreus* TM8 collected from a hot desert place (~50 °C) in South Egypt produced a new highly oxygenated tetracyclic meroterpenoid, terretonin M (**245**) (Figure 43). The crude extract of the mass mycelium and solid rice meida could slightly inhibit the growth of *Proteus* sp., *Candida albicans*, and *Streptococcus pyogenes*, while authors failed to isolate the active ingredient [80].

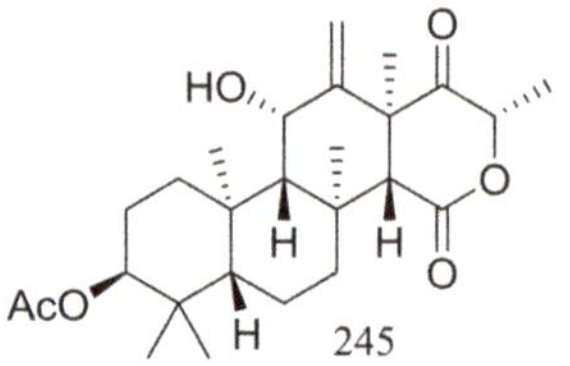

Figure 43. Novel natural products derived from thermophilic fungi (compound **245**).

5. Halophilic Fungi

Diverse novel compounds have been described from the halotolerant fungal strain *Aspergillus variecolor* B-17, which was isolated from the sediments collected in Jilantai salt field, Alashan, Inner Mongolia, China. Variecolorquinones A–B (**246–247**) (Figure 44) are two new quinone type compounds with cytotoxic activities against A549 cells (compound **246**, IC_{50} 3.0 μM), HL60 cells (compound **247**, IC_{50} 1.3 μM) and P388 cells (compound **247**, IC_{50} 3.7 μM) [81]. Variecolorins A–L (**248–259**) (Figure 44) exhibited no cytotoxicity (P388, HL-60, BEL-7402, and A-549 cell lines) but A–K (**248–258**) showed weak radical-scavenging activity against DPPH with IC_{50} values ranging from 75 to 102 μM [82]. Variecolortides A–C (**260–262**) (Figure 44) shared an unprecedented 'spiro-anthronopyranoid diketopiperazine' structure with a stable hemiaminal function. All three compounds showed weak cytotoxic activity against K-562 cell line with IC_{50} values of 61, 69 and 71 μM respectively (the positive control paclitaxel IC_{50} 0.93 μM) and showed very slight radical-scavenging activity against DPPH radical with IC$_{50}$ values of 63, 84 and 91 µM, respectively (the positive control vitamin C IC$_{50}$ 22 µM) [83].

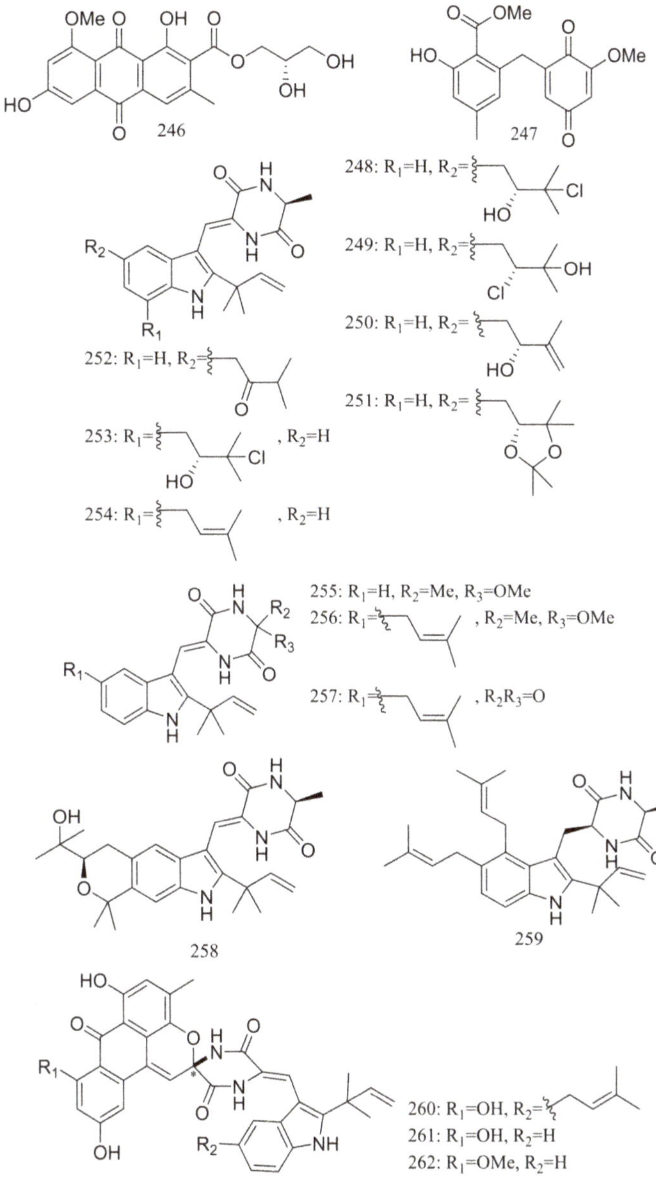

Figure 44. Novel natural products derived from halophilic fungi (compounds **246–262**). * Absolute configuration is not determined.

Pennicitrinone C (**263**) and penicitrinol B (**264**) (Figure 45), two new citrinin dimers were produced by the halotolerant fungal strain *Penicillium citrinum* B-57 obtained from the sediments in Jilantai salt field, Alashan, Inner Mongolia, China. Compound **263** scavenged DPPH radicals with IC$_{50}$ value of

55.3 µM (the positive control L-ascorbic acid IC_{50} 22.7 µM) but exhibited no cytotoxic activity against P388, A549, BEL7402 or HL60 cell lines (IC_{50} > 50 µM) [84].

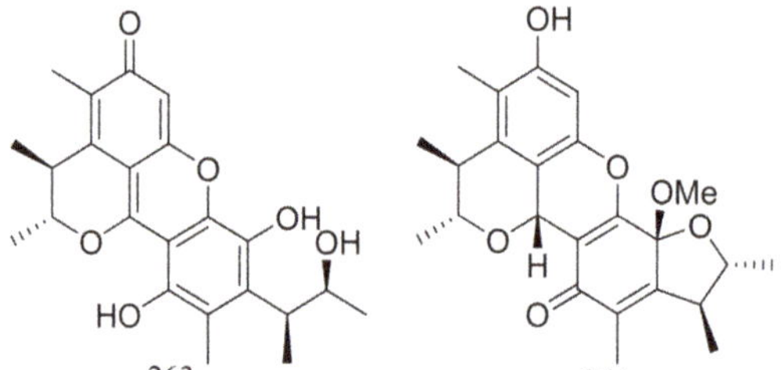

Figure 45. Novel natural products derived from halophilic fungi (compounds **263–264**).

Three new cerebrosides, alternarosides A–C (**265–267**) and one new diketopiperazine alkaloid, alternarosin A (**268**) (Figure 46) were produced by the halotolerant fungus *Alternaria raphani* THW-18, which was obtained from a sediment sample in the Hongdao sea salt field, China. Antimicrobial activities against *Escherichia coli*, *Bacillus subtilis*, and *Candida albicans* were evaluated and MIC values for four compounds ranged from 70 to 400 µM. Neither cytotoxicity (P388, HL-60, A549, and BEL-7402 cell lines) nor DPPH radical-scavenging activity was detected [85].

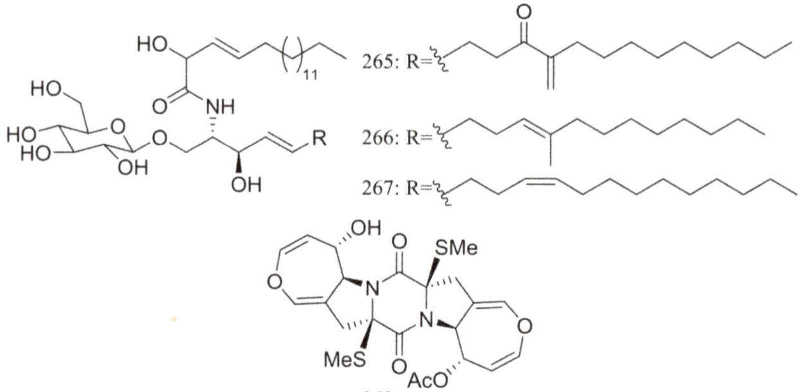

Figure 46. Novel natural products derived from halophilic fungi (compounds **265–268**).

Sclerotides A–B (**269–270**) (Figure 47) were novel cyclic hexapeptides produced by the halotolerant *Aspergillus sclerotiorum* PT06-1 (the Putian Sea Salt Field, China) in a nutrient-limited hypersaline medium. In the general bioactivity profiling programs including cytotoxic and antimicrobial testing, both compounds inhibited *Candida albicans* with MIC values of 7.0 and 3.5 µM respectively. Besides, compound **270** displayed weak cytotoxic activity against HL-60 cells (IC_{50} 56.1 µM) and antibacterial activity against *Pseudomonas aeruginosa* (MIC 35.3 µM) [86]. The same research group subsequently obtained eleven new aspochracin-type cyclic tripeptides named sclerotiotides A–K (**271–281**) (Figure 47) [87] and one new cytotoxic indole-3-ethenamide (**282**) (Figure 47) [88] from the same halotolerant fungal strain in a nutrient-rich hypersaline medium. Compounds **278–281** were four isomers with the same molecular formula, and the NMR data suggested that **278/280** and **279/281** were enantiotopic in the fatty acid moiety respectively. When evaluated for antimicrobial and cytotoxic activities compounds **271**, **272**, **276**, and **279** exhibited antifungal activities against *Candida albicans* with

MIC values of 7.5, 3.8, 30, and 6.7 µM respectively, while compound **282** showed cytotoxic activity against A-549 and HL-60 cells with IC$_{50}$ values of 3.0 and 27 µM respectively.

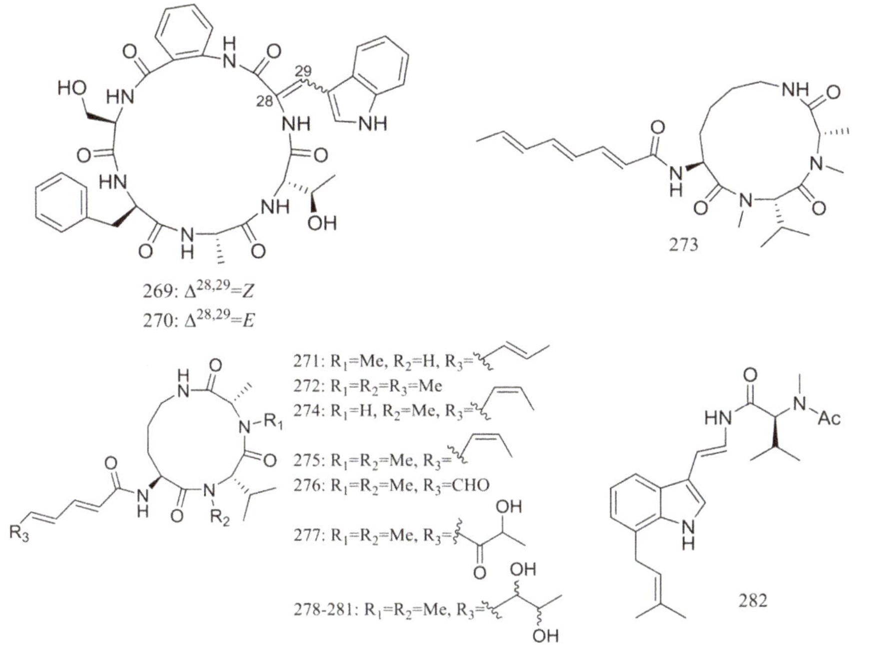

269: $\Delta^{28,29}=Z$
270: $\Delta^{28,29}=E$

271: R$_1$=Me, R$_2$=H, R$_3$=
272: R$_1$=R$_2$=R$_3$=Me
274: R$_1$=H, R$_2$=Me, R$_3$=
275: R$_1$=R$_2$=Me, R$_3$=
276: R$_1$=R$_2$=Me, R$_3$=CHO
277: R$_1$=R$_2$=Me, R$_3$=
278-281: R$_1$=R$_2$=Me, R$_3$=

273

282

Figure 47. Novel natural products derived from halophilic fungi (compounds **269–282**).

6. Xerophilic Fungi

Globosumones A–C (**283–285**) (Figure 48) are three new esters of orsellinic acid isolated from *Chaetomium globosum* endophytic on *Ephedra fasciculata* (Mormon tea), which was collected from the Sonoran Desert. Cytotoxic activities against four cancer cell lines (NCI-H460, MCF-7, SF-268, and MIA Pa Ca-2) were tested and only compounds **283** and **284** were moderately active with IC$_{50}$ values of 6.5 to 30.2 µM (the positive control doxorubicin IC$_{50}$ 0.01 to 0.07 µM) [89].

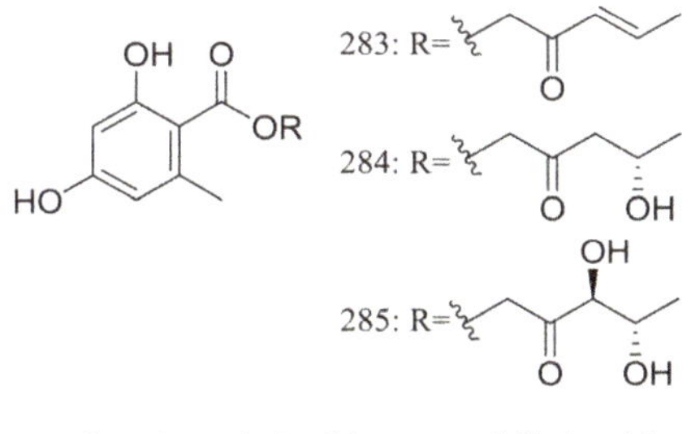

283: R=
284: R=
285: R=

Figure 48. Novel natural products derived from xerophilic fungi (compounds **283–285**).

The xerophilic fungus *Aspergillus restrictus* A-17 obtained from house dust yielded two new dioxopiperazine derivatives, arestrictins A–B (**286–287**) (Figure 49). The biological activity of them was not tested [90].

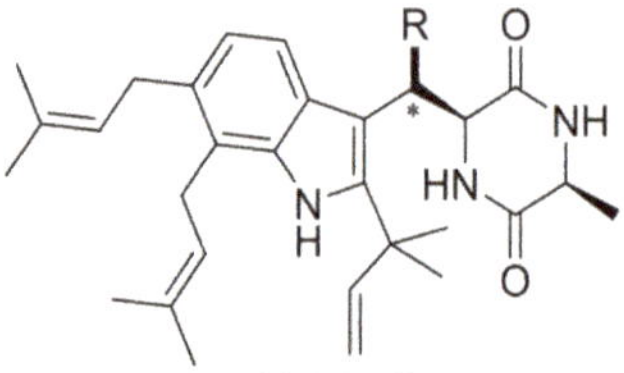

286: R=OH
287: R=H

Figure 49. Novel natural products derived from xerophilic fungi (compounds **286–287**). * Absolute configuration is not determined.

The culture broth of the volcanic ash-derived fungus *Penicillium citrinum* HGY1-5 collected from the extinct volcano Huguangyan in Guangdong, China, afforded eleven new unusual C25 steroid isomers with bicyclo[4.4.1]A/B rings named 24-*epi*-cyclocitrinol (**288**), 20-*O*-methyl-24-*epi*cyclocitrinol (**289**), 20-*O*-methylcyclocitrinol (**290**), 24-oxocyclocitrinol (**291**), 12*R*-hydroxycyclocitrinol (**292**), neocyclocitrinols B and D (**293** and **294**), *erythro*-23-*O*-methylneocyclocitrinol (**295**), *threo*-23-*O*-methylneocyclocitrinol (**296**), isocyclocitrinol B (**297**), and precyclocitrinol B (**298**) (Figure 50). The evaluation for biological activity of all steroids with the cAMP assay in GPR12-CHO and WT-CHO cells indicated that compounds **288**, **293** and **296** could induce the production of cAMP in GPR12-transfected CHO cells at 10 μM [91].

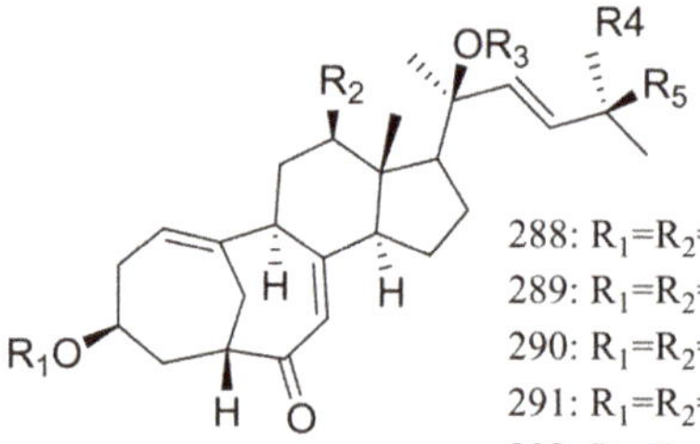

288: R_1=R_2=R_3=R_4=H, R_5=OH
289: R_1=R_2= R_4=H, R_3=Me, R_5=OH
290: R_1=R_2=R_5=H, R_3=Me, R_4=OH
291: R_1=R_2=R_3=H, R_4R_5=O
292: R_1=R_3=R_5=H, R_2=R_4=OH

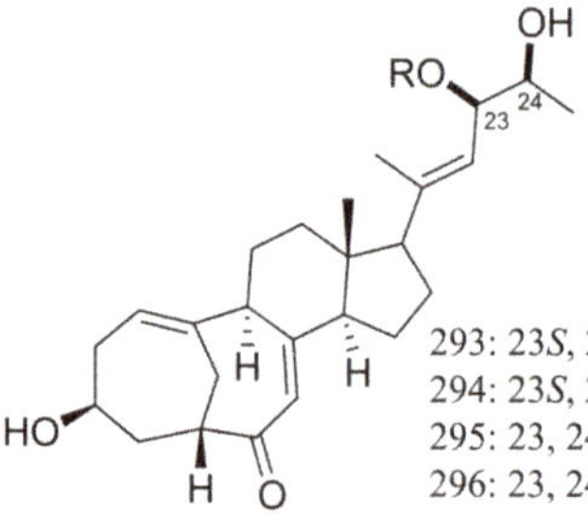

293: 23*S*, 24*S*, R=H
294: 23*S*, 24*R*, R=H
295: 23, 24-*erythro*, R=Me
296: 23, 24-*threo*, R=Me

Figure 50. *Cont.*

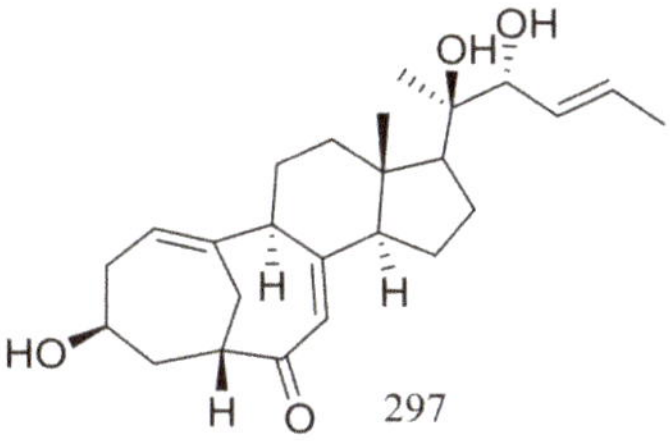

Figure 50. Novel natural products derived from xerophilic fungi (compounds **288–298**). * Absolute configuration is not determined.

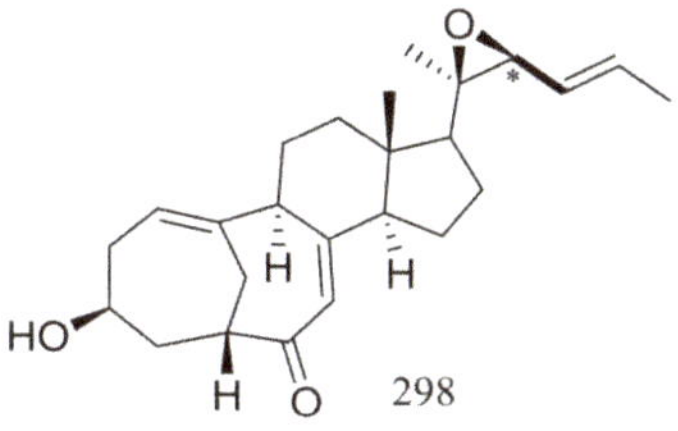

7. Acidophilic or Alkaliphilic Fungi

Since 2004, several new compounds have been obtained from *Penicillium* species growing in the Berkeley Pit Lake (Butte, Montana), which is an abandoned open-pit copper mine filled with 30 billion gallons of acidic, metal-contaminated water. Two novel hybrid polyketide-terpenoids named berkeleydione (**299**) and berkeleytrione (**300**) (Figure 51) [92], one novel spiroketal named berkelic acid (**301a**) (Figure 51) [93], as well as berkeleyacetals A–C (**302–304**) (Figure 51) [94] were isolated from an unidentified *Penicillium* species. In 2018 Fürstner group revised the absolute configuration of berkelic acid (**301b**) through an elegant synthetic, NMR, and crystallographic study [95]. Compounds **299–304** were found to be inhibitors of matrix metalloproteinase-3 (MMP-3) and the cysteine protease caspase-1 (Casp-1) in the micromolar or millimolar range. In addition, compounds **299** and **304** displayed cytotoxic activity against NCI-H460 cells (GI$_{50}$ 0.398 µM), while **301** against OVCAR-3 cells (GI$_{50}$ 0.091 µM). Berkeleyamides A–D (**305–308**) (Figure 51) [96] and berkeleyones A–C (**309–311**) (Figure 51) [97] were isolated from *Penicillium rubrum*. Compounds **305–308** were able to suppress caspase-1 and MMP-3 in the low micromolar range. Effects on inhibiting the production of interleukin 1-β in THP-1 cells was tested and IC$_{50}$ values for compounds **309** and **310** were 2.7 and 3.7 µM respectively (the positive control Ac-YVAD-CHO IC$_{50}$ 2.0 µM). Two new drimane sesquiterpene lactones named berkedrimanes A–B (**312–313**) and one new tricarboxylic acid derivative (**314**) (Figure 51) were produced by *Penicillium solitum*. Compounds **312** and **313** inhibited caspase-1 and caspase-3 in the micromolar range and mitigated the production of IL-1β by intact inflammasomes at low micromolar concentrations [98].

Table 1. Novel natural products isolated from extremophilic fungi.

The Type of Compound	Compounds	Biological Activity	References
Terpenoids and steroids	23, 24 *, 25–28	Cytotoxic	[25]
	(39–41) *, 42, 43 *, 44, 45	Cytotoxic and/or antiviral	[29,30]
	(145–150) *, 154, 155 *, 156, 157 *	Cytotoxic and/or antimicrobial	[52,53,56,57]
	176 *, 177–179, 180 *, 181–183, 184 *	Cytotoxic	[62,63]
	190, 191		[66]
	200 *, 201–204	Antiviral	[69]
	208		[70]
	209	Antiallergic	[71]
	245		[80]
	288 *, 289–292, 293 *, 294, 295, 296 *, 297, 298	Induce cAMP production	[91]
	299 *, 300 *	Inhibit MMP-3 and Casp-1 and/or cytotoxic	[92]
	309 *, 310 *, 311	Mitigate IL-1β production	[97]

Table 1. *Cont.*

The Type of Compound	Compounds	Biological Activity	References
Alkaloids, peptides, and amides	10, 11 *, 12, 13, 14 *	Anti-inflammatory	[24]
	15 *, 16, 17 *, 18 *, 19–22	Cytotoxic	[25,26]
	29–33		[27]
	(47–55) *	Insecticidal	[32]
	73		[36]
	74 *, 75, 76	Inhibit α-glucosidase	[37]
	86 *, 87 *, 88, 89	Antimicrobial	[40]
	104, 105		[43]
	108, 112		[44]
	(124–135) *, 136–138, (139–142) *	Antiviral or antimicrobial	[48,49]
	144 *	Cytotoxic	[51]
	151 *, 152, 153	Cytotoxic	[55]
	(167–169) *, 170–173	Cytotoxic	[60]
	185		[63]
	187–189		[65]
	192 *, 193 *, 194, 195, 196 *, 197	Reduce intracellular lipid accumulation	[67]
	198, 199 *	Antimicrobial	[68]
	221 *, 222 *, 223–226	Nematicidal	[74,75]
	233 *, 234 *	Cytotoxic	[77]
	(248–258) *, 259, (260–262) *	Radical-scavenging and/or cytotoxic	[82,83]
	(265–268) *	Antimicrobial	[85]
	(269–272) *, 273–275, 276 *, 277, 278, 279 *, 280, 281, 282 *	Antimicrobial and/or cytotoxic	[86–88]
	286, 287		[90]
	(305–308) *	Inhibit MMP-3 and Casp-1	[96]
	312 *, 313 *	Inhibit MMP-3 and Casp-1, and mitigate IL-1β production	[98]
Quinones and phenols	46		[31]
	56		[33]
	63–67, 68 *, 69–71	Antimicrobial and cytotoxic	[35]
	77	Cytotoxic	[38]
	79 *, 80 *, 81, 82 *, 83, 84 *, 85 *	Cytotoxic	[39]
	94–97, 98 *, 99	Activate Nrf2	[41]
	106 *, 107	Antilarval	[43]
	109, 110 *, 111 *	Inhibit BRD4	[44]
	114–119		[46]
	143 *	Antimicrobial	[50]
	158–160, 161 *, 162 *, 163–166	Antimicrobial	[58,59]
	246 *, 247 *	Cytotoxic	[81]
	263 *, 264	Radical-scavenging	[84]
Esters and lactones	78		[38]
	207 *	Inhibit PTP1B	[70]
	227 *, 228 *, 229–232	Nematicidal	[76]
	283 *, 284 *, 285	Cytotoxic	[89]
	(302–304) *	Cytotoxic and/or inhibit MMP-3	[94]
Xanthones	57 *, 58, 59 *, 60 *	Antimicrobial	[34]
	61, 62		[35]
	72		[36]
	90–93		[40]
Polyketides	100, 101, 102 *, 103 *	Cytotoxic	[42]
	120–122, 123 *	Cytotoxic	[47]
	174 *, 175	Inhibit NF-κB	[61]
	186 *	Antimicrobial	[64]
	(210–212) *, 213, 214	Cytotoxic	[72]
	215, 216, (217–220) *	Cytotoxic	[73]
	235–241, 242 *, 243 *	Cytotoxic	[78]

Table 1. Cont.

The Type of Compound	Compounds	Biological Activity	References
Others	(1–9) *	Cytotoxic	[20–23]
	(34–38) *	Antimicrobial	[28]
	113 *	Antimicrobial	[45]
	205 *, 206 *	Antiviral	[69]
	244 *	Antimicrobial	[79]
	301 *	Inhibit MMP-3 and Casp-1	[93]
	314		[98]

* bioactive compounds.

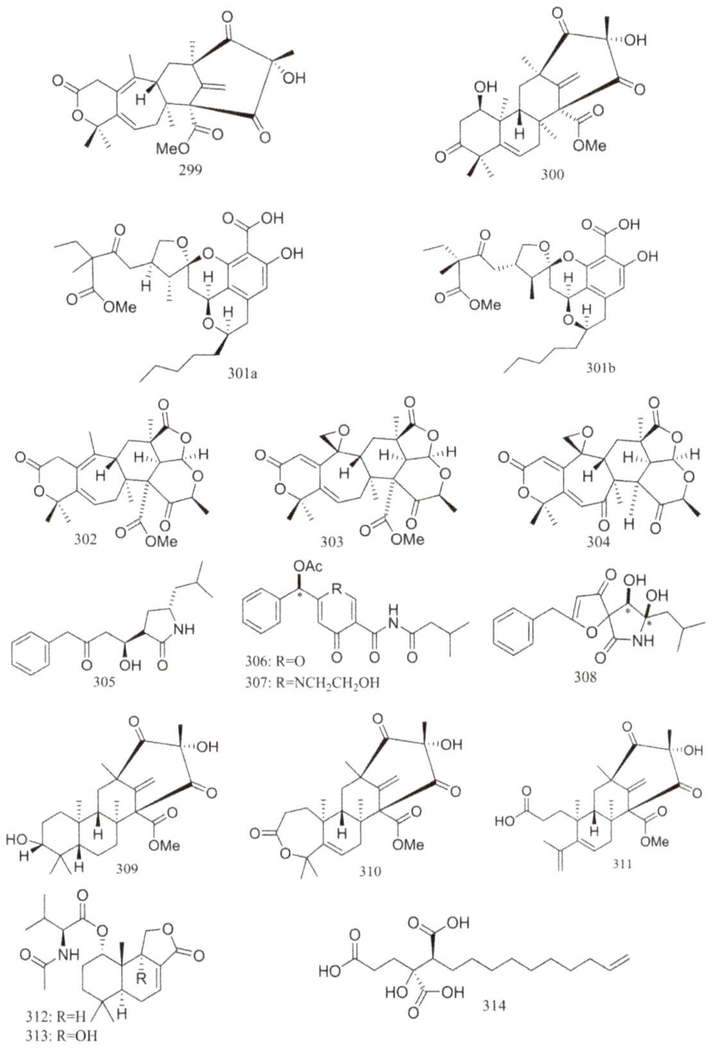

Figure 51. Novel natural products derived from acidophilic or alkaliphilic fungi (compounds **299–314**). * Absolute configuration is not determined.

8. Conclusions

In this review, a total of 314 novel compounds (161 bioactive ones) from extremophilic fungi have been compiled, including 58 terpenoids/steroids, 130 alkaloids/peptides/amides, 50 quinones/phenols, 14 esters/lactones, 11 xanthones, 31 polyketides, and 20 other structure compounds. All compounds were obtained from 56 fungal strains, most of which were asexual stages of ascomycetes e.g., *Penicillium* sp. (21 strains), *Aspergillus* sp. (11 strains), and other species (22 strains). Only one basidiomycete (*Acaromyces* sp.) and one zygomycete (*Malbranchea* sp.) appeared in the present review.

As demonstrated by this review, fungi from extreme environments are a rich source for novel natural products, even though the research on them is not as up-to-date as the research on fungi in other mesophilic environments due to the difficulties in both sample collection and cultivation. However, with the fast development of modern instruments and techniques in the post-genomic era, some groups have obtained many new compounds from one strain by changing its cultivation conditions or creating a mutant, which significantly contributes to make full use of these precious biological resources.

Author Contributions: X.Z. wrote the manuscript; S.-J.L., J.-J.L., Z.-Z.L., and C.-Q.Z. provided critical reviews and revisions of the manuscript.

Funding: This research was funded by the National Natural Sciences Foundation of China (NSFC) grant number [81173505].

Conflicts of Interest: The authors declare that they have no conflict of interests.

References

1. Macelroy, R.D. Some comments on the evolution of extremophiles. *BioSystem* **1974**, *6*, 74–75. [CrossRef]
2. Woese, C.R.; Kandler, O.; Wheelis, M.L. Towards a natural system of organisms: Proposal for the domains archaea, bacteria, and eucarya. *PNAS* **1990**, *87*, 4576–4579. [CrossRef] [PubMed]
3. Skropeta, D. Deep-sea natural products. *Nat. Prod. Rep.* **2008**, *25*, 1131–1166. [CrossRef] [PubMed]
4. Yayanos, A.A. Microbiology to 10,500 meters in the deep sea. *Annu. Rev. Microbiol.* **1995**, *49*, 777–805. [CrossRef] [PubMed]
5. Horikoshi, K. Barophiles: Deep-sea microorganisms adapted to an extreme environment. *Curr. Opin. Microbiol.* **1998**, *1*, 291–295. [CrossRef]
6. Wilson, Z.E.; Brimble, M.A. Molecules derived from the extremes of life. *Nat. Prod. Rep.* **2009**, *26*, 44–71. [CrossRef] [PubMed]
7. Deming, J.W. Psychrophiles and polar regions. *Curr. Opin. Microbiol.* **2002**, *5*, 301–309. [CrossRef]
8. Evans, R.D.; Johansen, J.R. Microbiotic crusts and ecosystem processes. *Crit. Rev. Plant Sci.* **1999**, *18*, 183–225. [CrossRef]
9. Stetter, K.O. Extremophiles and their adaptation to hot environments. *FEBS Lett.* **1999**, *452*, 22–25. [CrossRef]
10. Cavicchioli, R.; Thomas, T.; Curmi, P.M.G. Cold stress response in archaea. *Extremophiles* **2000**, *4*, 321–331. [CrossRef] [PubMed]
11. Madern, D.; Ebel, C.; Zaccai, G. Halophilic adaptation of enzymes. *Extremophiles* **2000**, *4*, 91–98. [CrossRef] [PubMed]
12. Rothschild, L.J.; Mancinelli, R.L. Life in extreme environments. *Nature* **2001**, *409*, 1092. [CrossRef] [PubMed]
13. Pakchung, A.A.H.; Simpson, P.J.L.; Codd, R. Life on earth. Extremophiles continue to move the goal posts. *Environ. Chem.* **2006**, *3*, 77–93. [CrossRef]
14. Lebar, M.D.; Heimbegner, J.L.; Baker, B.J. Cold-water marine natural products. *Nat. Prod. Rep.* **2007**, *24*, 774–797. [CrossRef] [PubMed]
15. Soldatou, S.; Baker, B.J. Cold-water marine natural products, 2006 to 2016. *Nat. Prod. Rep.* **2017**, *34*, 585–626. [CrossRef] [PubMed]
16. Jing-Tang, L.; Xiao-Ling, L.; Xiao-Yu, L.; Yun, G.; Bo, H.; Bing-Hua, J.; Heng, Z. Bioactive natural products from the antarctic and arctic organisms. *Mini-Rev. Med. Chem.* **2013**, *13*, 617–626. [CrossRef]

17. Tian, Y.; Li, Y.-L.; Zhao, F.-C. Secondary metabolites from polar organisms. *Mar. Drugs* **2017**, *15*. [CrossRef] [PubMed]

18. Skropeta, D.; Wei, L. Recent advances in deep-sea natural products. *Nat. Prod. Rep.* **2014**, *31*, 999–1025. [CrossRef] [PubMed]

19. Wang, Y.-T.; Xue, Y.-R.; Liu, C.-H. A brief review of bioactive metabolites derived from deep-sea fungi. *Mar. Drugs* **2015**, *13*. [CrossRef] [PubMed]

20. Li, D.; Wang, F.; Cai, S.; Zeng, X.; Xiao, X.; Gu, Q.; Zhu, W. Two new bisorbicillinoids isolated from a deep-sea fungus, *Phialocephala* sp. FL30r. *J. Antibiot.* **2007**, *60*, 317–320. [CrossRef] [PubMed]

21. Li, D.; Wang, F.; Xiao, X.; Fang, Y.; Zhu, T.; Gu, Q.; Zhu, W. Trisorbicillinone a, a novel sorbicillin trimer, from a deep sea fungus, *Phialocephala* sp. FL30r. *Tetrahedron Lett.* **2007**, *48*, 5235–5238. [CrossRef]

22. Li, D.; Cai, S.; Zhu, T.; Wang, F.; Xiao, X.; Gu, Q. Three new sorbicillin trimers, trisorbicillinones b, c, and d, from a deep ocean sediment derived fungus, *Phialocephala* sp. FL30r. *Tetrahedron* **2010**, *66*, 5101–5106. [CrossRef]

23. Li, D.-H.; Cai, S.-X.; Zhu, T.-J.; Wang, F.-P.; Xiao, X.; Gu, Q.-Q. New cytotoxic metabolites from a deep-sea-derived fungus, *Phialocephala* sp., strain FL30r. *Chem. Biodivers.* **2011**, *8*, 895–901. [CrossRef] [PubMed]

24. Du, L.; Yang, X.; Zhu, T.; Wang, F.; Xiao, X.; Park, H.; Gu, Q. Diketopiperazine alkaloids from a deep ocean sediment derived fungus *Penicillium* sp. *Chem. Pharm. Bull.* **2009**, *57*, 873–876. [CrossRef] [PubMed]

25. Du, L.; Li, D.; Zhu, T.; Cai, S.; Wang, F.; Xiao, X.; Gu, Q. New alkaloids and diterpenes from a deep ocean sediment derived fungus *Penicillium* sp. *Tetrahedron* **2009**, *65*, 1033–1039. [CrossRef]

26. Du, L.; Feng, T.; Zhao, B.; Li, D.; Cai, S.; Zhu, T.; Wang, F.; Xiao, X.; Gu, Q. Alkaloids from a deep ocean sediment-derived fungus *Penicillium* sp. and their antitumor activities. *J. Antibiot.* **2010**, *63*, 165–170. [CrossRef] [PubMed]

27. Guo, W.; Peng, J.; Zhu, T.; Gu, Q.; Keyzers, R.A.; Li, D. Sorbicillamines a–e, nitrogen-containing sorbicillinoids from the deep-sea-derived fungus *Penicillium* sp. F23-2. *J. Nat. Prod.* **2013**, *76*, 2106–2112. [CrossRef] [PubMed]

28. Guo, W.; Zhang, Z.; Zhu, T.; Gu, Q.; Li, D. Penicyclones a–e, antibacterial polyketides from the deep-sea-derived fungus *Penicillium* sp. F23-2. *J. Nat. Prod.* **2015**, *78*, 2699–2703. [CrossRef] [PubMed]

29. Li, Y.; Ye, D.; Chen, X.; Lu, X.; Shao, Z.; Zhang, H.; Che, Y. Breviane spiroditerpenoids from an extreme-tolerant *Penicillium* sp. isolated from a deep sea sediment sample. *J. Nat. Prod.* **2009**, *72*, 912–916. [CrossRef] [PubMed]

30. Li, Y.; Ye, D.; Shao, Z.; Cui, C.; Che, Y. A sterol and spiroditerpenoids from a *Penicillium* sp. isolated from a deep sea sediment sample. *Mar. Drugs* **2012**, *10*. [CrossRef] [PubMed]

31. Tian, Y.-Q.; Lin, X.-P.; Liu, J.; Kaliyaperumal, K.; Ai, W.; Ju, Z.-R.; Yang, B.; Wang, J.; Yang, X.-W.; Liu, Y. Ascomycotin a, a new citromycetin analogue produced by *Ascomycota* sp. Ind19F07 isolated from deep sea sediment. *Nat. Prod. Res.* **2015**, *29*, 820–826. [CrossRef] [PubMed]

32. Xu, X.; Zhang, X.; Nong, X.; Wei, X.; Qi, S. Oxindole alkaloids from the fungus *Penicillium commune* DFFSCS026 isolated from deep-sea-derived sediments. *Tetrahedron* **2015**, *71*, 610–615. [CrossRef]

33. Fredimoses, M.; Zhou, X.; Ai, W.; Tian, X.; Yang, B.; Lin, X.; Xian, J.-Y.; Liu, Y. Westerdijkin a, a new hydroxyphenylacetic acid derivative from deep sea fungus *Aspergillus westerdijkiae* SCSIO 05233. *Nat. Prod. Res.* **2015**, *29*, 158–162. [CrossRef] [PubMed]

34. Fredimoses, M.; Zhou, X.; Lin, X.; Tian, X.; Ai, W.; Wang, J.; Liao, S.; Liu, J.; Yang, B.; Yang, X.; et al. New prenylxanthones from the deep-sea derived fungus *Emericella* sp. SCSIO 05240. *Mar. Drugs* **2014**, *12*. [CrossRef] [PubMed]

35. Yao, Q.; Wang, J.; Zhang, X.; Nong, X.; Xu, X.; Qi, S. Cytotoxic polyketides from the deep-sea-derived fungus *Engyodontium album* DFFSCS021. *Mar. Drugs* **2014**, *12*. [CrossRef] [PubMed]

36. Tian, Y.; Qin, X.; Lin, X.; Kaliyaperumal, K.; Zhou, X.; Liu, J.; Ju, Z.; Tu, Z.; Liu, Y. Sydoxanthone c and acremolin b produced by deep-sea-derived fungus *Aspergillus* sp. SCSIO Ind09F01. *J. Antibiot.* **2015**, *68*, 703–706. [CrossRef] [PubMed]

37. Fan, Z.; Sun, Z.-H.; Liu, Z.; Chen, Y.-C.; Liu, H.-X.; Li, H.-H.; Zhang, W.-M. Dichotocejpins a–c: New diketopiperazines from a deep-sea-derived fungus *Dichotomomyces cejpii* FS110. *Mar. Drugs* **2016**, *14*. [CrossRef] [PubMed]

38. Gao, X.-W.; Liu, H.-X.; Sun, Z.-H.; Chen, Y.-C.; Tan, Y.-Z.; Zhang, W.-M. Secondary metabolites from the deep-sea derived fungus *Acaromyces ingoldii* FS121. *Molecules* **2016**, *21*. [CrossRef]

39. Zhang, Z.; Guo, W.; He, X.; Che, Q.; Zhu, T.; Gu, Q.; Li, D. Peniphenylanes a–g from the deep-sea-derived fungus *Penicillium fellutanum* HDN14-323. *Planta Med.* **2016**, *82*, 872–876. [CrossRef] [PubMed]

40. Wang, J.; He, W.; Huang, X.; Tian, X.; Liao, S.; Yang, B.; Wang, F.; Zhou, X.; Liu, Y. Antifungal new oxepine-containing alkaloids and xanthones from the deep-sea-derived fungus *Aspergillus versicolor* SCSIO 05879. *J. Agric. Food Chem.* **2016**, *64*, 2910–2916. [CrossRef] [PubMed]

41. Wu, Z.; Wang, Y.; Liu, D.; Proksch, P.; Yu, S.; Lin, W. Antioxidative phenolic compounds from a marine-derived fungus *Aspergillus versicolor*. *Tetrahedron* **2016**, *72*, 50–57. [CrossRef]

42. Zhang, Z.; He, X.; Liu, C.; Che, Q.; Zhu, T.; Gu, Q.; Li, D. Clindanones a and b and cladosporols f and g, polyketides from the deep-sea derived fungus *Cladosporium cladosporioides* HDN14-342. *RSC Adv.* **2016**, *6*, 76498–76504. [CrossRef]

43. Xu, X.; Zhang, X.; Nong, X.; Wang, J.; Qi, S. Brevianamides and mycophenolic acid derivatives from the deep-sea-derived fungus *Penicillium brevicompactum* DFFSCS025. *Mar. Drugs* **2017**, *15*. [CrossRef] [PubMed]

44. Ding, H.; Zhang, D.; Zhou, B.; Ma, Z. Inhibitors of BRD4 protein from a marine-derived fungus *Alternaria* sp. NH-F6. *Mar. Drugs* **2017**, *15*. [CrossRef] [PubMed]

45. Wang, W.; Li, S.; Chen, Z.; Li, Z.; Liao, Y.; Chen, J. Secondary metabolites produced by the deep-sea-derived fungus *Engyodontium album*. *Chem. Nat. Compd.* **2017**, *53*, 224–226. [CrossRef]

46. Luo, X.; Lin, X.; Salendra, L.; Pang, X.; Dai, Y.; Yang, B.; Liu, J.; Wang, J.; Zhou, X.; Liu, Y. Isobenzofuranones and isochromenones from the deep-sea derived fungus *Leptosphaeria* sp. SCSIO 41005. *Mar. Drugs* **2017**, *15*. [CrossRef] [PubMed]

47. Zhang, Z.; He, X.; Zhang, G.; Che, Q.; Zhu, T.; Gu, Q.; Li, D. Inducing secondary metabolite production by combined culture of *Talaromyces aculeatus* and *Penicillium variabile*. *J. Nat. Prod.* **2017**, *80*, 3167–3171. [CrossRef] [PubMed]

48. Niu, S.; Liu, D.; Shao, Z.; Proksch, P.; Lin, W. Eutypellazines a-m, thiodiketopiperazine-type alkaloids from deep sea derived fungus *Eutypella* sp. MCCC 3A00281. *RSC Adv.* **2017**, *7*, 33580–33590. [CrossRef]

49. Niu, S.; Liu, D.; Shao, Z.; Proksch, P.; Lin, W. Eutypellazines n−s, new thiodiketopiperazines from a deep sea sediment derived fungus *Eutypella* sp. with anti-VRE activities. *Tetrahedron Lett.* **2017**, *58*, 3695–3699. [CrossRef]

50. Takahashi, K.; Sakai, K.; Nagano, Y.; Orui Sakaguchi, S.; Lima, A.O.; Pellizari, V.H.; Iwatsuki, M.; Takishita, K.; Nonaka, K.; Fujikura, K.; et al. Cladomarine, a new anti-saprolegniasis compound isolated from the deep-sea fungus, *Penicillium coralligerum* YK-247. *J. Antibiot.* **2017**, *70*, 911–914. [CrossRef] [PubMed]

51. Dalsgaard, P.W.; Larsen, T.O.; Christophersen, C. Bioactive cyclic peptides from the psychrotolerant fungus *Penicillium algidum*. *J. Antibiot.* **2005**, *58*, 141–144. [CrossRef] [PubMed]

52. Oh, D.-C.; Jensen, P.R.; Kauffman, C.A.; Fenical, W. Libertellenones a–d: Induction of cytotoxic diterpenoid biosynthesis by marine microbial competition. *Biorg. Med. Chem.* **2005**, *13*, 5267–5273. [CrossRef] [PubMed]

53. Lu, X.-L.; Liu, J.-T.; Liu, X.-Y.; Gao, Y.; Zhang, J.; Jiao, B.-H.; Zheng, H. Pimarane diterpenes from the arctic fungus *Eutypella* sp. D-1. *J. Antibiot.* **2014**, *67*, 171–174. [CrossRef] [PubMed]

54. Shen, C.; Xu, N.; Gao, Y.; Sun, X.; Yin, Y.; Cai, M.; Zhou, X.; Zhang, Y. Stimulatory effect of ethanol on libertellenone h biosynthesis by arctic fungus *Eutypella* sp. D-1. *Bioprocess Biosyst. Eng.* **2016**, *39*, 353–360. [CrossRef] [PubMed]

55. Liu, J.-T.; Hu, B.; Gao, Y.; Zhang, J.-P.; Jiao, B.-H.; Lu, X.-L.; Liu, X.-Y. Bioactive tyrosine-derived cytochalasins from fungus *Eutypella* sp. D-1. *Chem. Biodivers.* **2014**, *11*, 800–806. [CrossRef] [PubMed]

56. Zhang, L.-Q.; Chen, X.-C.; Chen, Z.-Q.; Wang, G.-M.; Zhu, S.-G.; Yang, Y.-F.; Chen, K.-X.; Liu, X.-Y.; Li, Y.-M. Eutypenoids a–c: Novel pimarane diterpenoids from the arctic fungus *Eutypella* sp. D-1. *Mar. Drugs* **2016**, *14*. [CrossRef] [PubMed]

57. Zhou, Y.; Zhang, Y.-X.; Zhang, J.-P.; Yu, H.-B.; Liu, X.-Y.; Lu, X.-L.; Jiao, B.-H. A new sesquiterpene lactone from fungus *Eutypella* sp. D-1. *Nat. Prod. Res.* **2017**, *31*, 1676–1681. [CrossRef] [PubMed]

58. Li, Y.; Sun, B.; Liu, S.; Jiang, L.; Liu, X.; Zhang, H.; Che, Y. Bioactive asterric acid derivatives from the antarctic ascomycete fungus *Geomyces* sp. *J. Nat. Prod.* **2008**, *71*, 1643–1646. [CrossRef] [PubMed]

59. Figueroa, L.; Jiménez, C.; Rodríguez, J.; Areche, C.; Chávez, R.; Henríquez, M.; de la Cruz, M.; Díaz, C.; Segade, Y.; Vaca, I. 3-nitroasterric acid derivatives from an antarctic sponge-derived *Pseudogymnoascus* sp. Fungus. *J. Nat. Prod.* **2015**, *78*, 919–923. [CrossRef] [PubMed]

60. Li, L.; Li, D.; Luan, Y.; Gu, Q.; Zhu, T. Cytotoxic metabolites from the antarctic psychrophilic fungus *Oidiodendron truncatum*. *J. Nat. Prod.* **2012**, *75*, 920–927. [CrossRef] [PubMed]

61. Wu, G.; Ma, H.; Zhu, T.; Li, J.; Gu, Q.; Li, D. Penilactones a and b, two novel polyketides from antarctic deep-sea derived fungus *Penicillium crustosum* PRB-2. *Tetrahedron* **2012**, *68*, 9745–9749. [CrossRef]

62. Wu, G.; Lin, A.; Gu, Q.; Zhu, T.; Li, D. Four new chloro-eremophilane sesquiterpenes from an antarctic deep-sea derived fungus, *Penicillium* sp. PR19N-1. *Mar. Drugs* **2013**, *11*. [CrossRef] [PubMed]

63. Lin, A.; Wu, G.; Gu, Q.; Zhu, T.; Li, D. New eremophilane-type sesquiterpenes from an antarctic deep-sea derived fungus, *Penicillium* sp. PR19N-1. *Arch. Pharm. Res.* **2014**, *37*, 839–844. [CrossRef] [PubMed]

64. Wu, B.; Wiese, J.; Labes, A.; Kramer, A.; Schmaljohann, R.; Imhoff, J.F. Lindgomycin, an unusual antibiotic polyketide from a marine fungus of the *Lindgomycetaceae*. *Mar. Drugs* **2015**, *13*, 4617–4632. [CrossRef] [PubMed]

65. Wang, J.; He, W.; Qin, X.; Wei, X.; Tian, X.; Liao, L.; Liao, S.; Yang, B.; Tu, Z.; Chen, B.; et al. Three new indolyl diketopiperazine metabolites from the antarctic soil-derived fungus *Penicillium* sp. SCSIO 05705. *RSC Adv.* **2015**, *5*, 68736–68742. [CrossRef]

66. Zhou, H.; Li, L.; Wang, W.; Che, Q.; Li, D.; Gu, Q.; Zhu, T. Chrodrimanins i and j from the antarctic moss-derived fungus *Penicillium funiculosum* GWT2-24. *J. Nat. Prod.* **2015**, *78*, 1442–1445. [CrossRef] [PubMed]

67. Zhou, H.; Li, L.; Wu, C.; Kurtán, T.; Mándi, A.; Liu, Y.; Gu, Q.; Zhu, T.; Guo, P.; Li, D. Penipyridones a–f, pyridone alkaloids from *Penicillium funiculosum*. *J. Nat. Prod.* **2016**, *79*, 1783–1790. [CrossRef] [PubMed]

68. Zhang, T.; Zhu, M.-L.; Sun, G.-Y.; Li, N.; Gu, Q.-Q.; Li, D.-H.; Che, Q.; Zhu, T.-J. Exopisiod b and farylhydrazone c, two new alkaloids from the antarctic-derived fungus *Penicillium* sp. HDN14-431. *J. Asian Nat. Prod. Res.* **2016**, *18*, 959–965. [CrossRef] [PubMed]

69. Wang, J.; Wei, X.; Qin, X.; Tian, X.; Liao, L.; Li, K.; Zhou, X.; Yang, X.; Wang, F.; Zhang, T.; et al. Antiviral merosesquiterpenoids produced by the antarctic fungus *Aspergillus ochraceopetaliformis* SCSIO 05702. *J. Nat. Prod.* **2016**, *79*, 59–65. [CrossRef] [PubMed]

70. Zhou, Y.; Li, Y.-H.; Yu, H.-B.; Liu, X.-Y.; Lu, X.-L.; Jiao, B.-H. Furanone derivative and sesquiterpene from antarctic marine-derived fungus *Penicillium sp. S-1-18*. *J. Asian Nat. Prod. Res.* **2017**, 1–8. [CrossRef] [PubMed]

71. Niu, S.; Fan, Z.-W.; Xie, C.-L.; Liu, Q.; Luo, Z.-H.; Liu, G.; Yang, X.-W. Spirograterpene a, a tetracyclic spiro-diterpene with a fused 5/5/5/5 ring system from the deep-sea-derived fungus *Penicillium granulatum* MCCC 3A00475. *J. Nat. Prod.* **2017**, *80*, 2174–2177. [CrossRef] [PubMed]

72. Yang, Y.L.; Lu, C.P.; Chen, M.Y.; Chen, K.Y.; Wu, Y.C.; Wu, S.H. Cytotoxic polyketides containing tetramic acid moieties isolated from the fungus *Myceliophthora Thermophila*: Elucidation of the relationship between cytotoxicity and stereoconfiguration. *Chem. Eur. J.* **2007**, *13*, 6985–6991. [CrossRef] [PubMed]

73. Yang, Y.-L.; Liao, W.-Y.; Liu, W.-Y.; Liaw, C.-C.; Shen, C.-N.; Huang, Z.-Y.; Wu, S.-H. Discovery of new natural products by intact-cell mass spectrometry and LC-SPE-NMR: Malbranpyrroles, novel polyketides from thermophilic fungus *Malbranchea sulfurea*. *Chem. Eur. J.* **2009**, *15*, 11573–11580. [CrossRef] [PubMed]

74. Chu, Y.-S.; Niu, X.-M.; Wang, Y.-L.; Guo, J.-P.; Pan, W.-Z.; Huang, X.-W.; Zhang, K.-Q. Isolation of putative biosynthetic intermediates of prenylated indole alkaloids from a thermophilic fungus *Talaromyces thermophilus*. *Org. Lett.* **2010**, *12*, 4356–4359. [CrossRef] [PubMed]

75. Guo, J.-P.; Tan, J.-L.; Wang, Y.-L.; Wu, H.-Y.; Zhang, C.-P.; Niu, X.-M.; Pan, W.-Z.; Huang, X.-W.; Zhang, K.-Q. Isolation of talathermophilins from the thermophilic fungus *Talaromyces thermophilus* YM3-4. *J. Nat. Prod.* **2011**, *74*, 2278–2281. [CrossRef] [PubMed]

76. Guo, J.-P.; Zhu, C.-Y.; Zhang, C.-P.; Chu, Y.-S.; Wang, Y.-L.; Zhang, J.-X.; Wu, D.-K.; Zhang, K.-Q.; Niu, X.-M. Thermolides, potent nematocidal pks-nrps hybrid metabolites from thermophilic fungus *Talaromyces thermophilus*. *J. Am. Chem. Soc.* **2012**, *134*, 20306–20309. [CrossRef] [PubMed]

77. Jiang, W.; Ye, P.; Chen, A.C.-T.; Wang, K.; Liu, P.; He, S.; Wu, X.; Gan, L.; Ye, Y.; Wu, B. Two novel hepatocellular carcinoma cycle inhibitory cyclodepsipeptides from a hydrothermal vent crab-associated fungus *Aspergillus clavatus* C2WU. *Mar. Drugs* **2013**, *11*. [CrossRef] [PubMed]

78. Chen, X.-W.; Li, C.-W.; Cui, C.-B.; Hua, W.; Zhu, T.-J.; Gu, Q.-Q. Nine new and five known polyketides derived from a deep sea-sourced *Aspergillus* sp. 16-02-1. *Mar. Drugs* **2014**, *12*. [CrossRef] [PubMed]

79. Pan, C.; Shi, Y.; Auckloo, N.B.; Chen, X.; Chen, A.C.-T.; Tao, X.; Wu, B. An unusual conformational isomer of verrucosidin backbone from a hydrothermal vent fungus, *Penicillium* sp. Y-50-10. *Mar. Drugs* **2016**, *14*. [CrossRef] [PubMed]

80. Shaaban, M.; El-Metwally, M.M.; Abdel-Razek, A.A.; Laatsch, H. Terretonin m: A new meroterpenoid from the thermophilic *Aspergillus terreus* TM8 and revision of the absolute configuration of penisimplicins. *Nat. Prod. Res.* **2017**, 1–10. [CrossRef] [PubMed]

81. Wang, W.; Zhu, T.; Tao, H.; Lu, Z.; Fang, Y.; Gu, Q.; Zhu, W. Two new cytotoxic quinone type compounds from the halotolerant fungus *Aspergillus variecolor*. *J. Antibiot.* **2007**, *60*, 603. [CrossRef] [PubMed]

82. Wang, W.-L.; Lu, Z.-Y.; Tao, H.-W.; Zhu, T.-J.; Fang, Y.-C.; Gu, Q.-Q.; Zhu, W.-M. Isoechinulin-type alkaloids, variecolorins a–l, from halotolerant *Aspergillus variecolor*. *J. Nat. Prod.* **2007**, *70*, 1558–1564. [CrossRef] [PubMed]

83. Wang, W.-L.; Zhu, T.-J.; Tao, H.-W.; Lu, Z.-Y.; Fang, Y.-C.; Gu, Q.-Q.; Zhu, W.-M. Three novel, structurally unique spirocyclic alkaloids from the halotolerant B-17 fungal strain of *Aspergillus variecolor*. *Chem. Biodivers.* **2007**, *4*, 2913–2919. [CrossRef] [PubMed]

84. Lu, Z.-Y.; Lin, Z.-J.; Wang, W.-L.; Du, L.; Zhu, T.-J.; Fang, Y.-C.; Gu, Q.-Q.; Zhu, W.-M. Citrinin dimers from the halotolerant fungus *Penicillium citrinum* B-57. *J. Nat. Prod.* **2008**, *71*, 543–546. [CrossRef] [PubMed]

85. Wang, W.; Wang, Y.; Tao, H.; Peng, X.; Liu, P.; Zhu, W. Cerebrosides of the halotolerant fungus *Alternaria raphani* isolated from a sea salt field. *J. Nat. Prod.* **2009**, *72*, 1695–1698. [CrossRef] [PubMed]

86. Zheng, J.; Zhu, H.; Hong, K.; Wang, Y.; Liu, P.; Wang, X.; Peng, X.; Zhu, W. Novel cyclic hexapeptides from marine-derived fungus, *Aspergillus sclerotiorum* PT06-1. *Org. Lett.* **2009**, *11*, 5262–5265. [CrossRef] [PubMed]

87. Zheng, J.; Xu, Z.; Wang, Y.; Hong, K.; Liu, P.; Zhu, W. Cyclic tripeptides from the halotolerant fungus *Aspergillus sclerotiorum* PT06-1. *J. Nat. Prod.* **2010**, *73*, 1133–1137. [CrossRef] [PubMed]

88. Wang, H.; Zheng, J.-K.; Qu, H.-J.; Liu, P.-P.; Wang, Y.; Zhu, W.-M. A new cytotoxic indole-3-ethenamide from the halotolerant fungus *Aspergillus sclerotiorum* PT06-1. *J. Antibiot.* **2011**, *64*, 679. [CrossRef] [PubMed]

89. Bashyal, B.P.; Wijeratne, E.M.K.; Faeth, S.H.; Gunatilaka, A.A.L. Globosumones a−c, cytotoxic orsellinic acid esters from the sonoran desert endophytic fungus *Chaetomium globosum*. *J. Nat. Prod.* **2005**, *68*, 724–728. [CrossRef] [PubMed]

90. Itabashi, T.; Matsuishi, N.; Hosoe, T.; Toyazaki, N.; Udagawa, S.; Imai, T.; Adachi, M.; Kawai, K. Two new dioxopiperazine derivatives, arestrictins a and b, isolated from *Aspergillus restrictus* and *Aspergillus penicilloides*. *Chem. Pharm. Bull.* **2006**, *54*, 1639. [CrossRef] [PubMed]

91. Du, L.; Zhu, T.; Fang, Y.; Gu, Q.; Zhu, W. Unusual C25 steroid isomers with bicyclo[4.4.1]a/b rings from a volcano ash-derived fungus *Penicillium citrinum*. *J. Nat. Prod.* **2008**, *71*, 1343–1351. [CrossRef] [PubMed]

92. Stierle, D.B.; Stierle, A.A.; Hobbs, J.D.; Stokken, J.; Clardy, J. Berkeleydione and berkeleytrione, new bioactive metabolites from an acid mine organism. *Org. Lett.* **2004**, *6*, 1049–1052. [CrossRef] [PubMed]

93. Stierle, A.A.; Stierle, D.B.; Kelly, K. Berkelic acid, a novel spiroketal with selective anticancer activity from an acid mine waste fungal extremophile. *J. Org. Chem.* **2006**, *71*, 5357–5360. [CrossRef] [PubMed]

94. Stierle, D.B.; Stierle, A.A.; Patacini, B. The berkeleyacetals, three meroterpenes from a deep water acid mine waste *Penicillium*. *J. Nat. Prod.* **2007**, *70*, 1820–1823. [CrossRef] [PubMed]

95. Bender, C.F.; Paradise, C.L.; Lynch, V.M.; Yoshimoto, F.K.; De Brabander, J.K. A biosynthetically inspired synthesis of (−)-berkelic acid and analogs. *Tetrahedron* **2018**, *74*, 909–919. [CrossRef]

96. Stierle, A.A.; Stierle, D.B.; Patacini, B. The berkeleyamides, amides from the acid lake fungus *Penicillum rubrum*. *J. Nat. Prod.* **2008**, *71*, 856–860. [CrossRef] [PubMed]

97. Stierle, D.B.; Stierle, A.A.; Patacini, B.; McIntyre, K.; Girtsman, T.; Bolstad, E. Berkeleyones and related meroterpenes from a deep water acid mine waste fungus that inhibit the production of interleukin 1-β from induced inflammasomes. *J. Nat. Prod.* **2011**, *74*, 2273–2277. [CrossRef] [PubMed]

98. Stierle, D.B.; Stierle, A.A.; Girtsman, T.; McIntyre, K.; Nichols, J. Caspase-1 and -3 inhibiting drimane sesquiterpenoids from the extremophilic fungus *Penicillium solitum*. *J. Nat. Prod.* **2012**, *75*, 262–266. [CrossRef] [PubMed]

 *marine drugs*

Article

Highly Substituted Benzophenone Aldehydes and Eremophilane Derivatives from the Deep-Sea Derived Fungus *Phomopsis lithocarpus* FS508

Jian-Lin Xu [1,2,†], Hong-Xin Liu [1,3,†], Yu-Chan Chen [1], Hai-Bo Tan [3], Heng Guo [1,2], Li-Qiong Xu [1], Sai-Ni Li [1], Zi-Lei Huang [1], Hao-Hua Li [1], Xiao-Xia Gao [2,*] and Wei-Min Zhang [1,*]

[1]   State Key Laboratory of Applied Microbiology Southern China, Guangdong Provincial Key Laboratory of Microbial Culture Collection and Application, Guangdong Open Laboratory of Applied Microbiology, Guangdong Institute of Microbiology, Guangzhou 510070, China; 15876503402@163.com (J.-L.X.); liuhx@gdim.cn (H.-X.L.); chenyc@gdim.cn (Y.-C.C.); Hengguo163@163.com (H.G.); xlq9513@126.com (L.-Q.X.); maibao66@126.com (S.-N.L.); huangzilei15@mails.ucas.ac.cn (Z.-L.H.); lihh@gdim.cn (H.-H.L.)

[2]   College of Pharmacy, Guangdong Pharmaceutical University, Guangzhou 510006, China

[3]   Program for Natural Products Chemical Biology, Key Laboratory of Plant Resources Conservation and Sustainable Utilization, Guangdong Provincial Key Laboratory of Applied Botany, South China Botanical Garden, Chinese Academy of Sciences, Guangzhou 510650, China; tanhaibo@scbg.ac.cn

*   Correspondence: gaoxxia91@163.com (X.-X.G.); wmzhang@gdim.cn (W.-M.Z.); Tel.: +86-20-8768-8309 (W.-M.Z.)

†   These authors contributed equally to this work.

Received: 10 August 2018; Accepted: 7 September 2018; Published: 11 September 2018

Abstract: Five new benzophenone derivatives named tenellones D–H (**1–5**), sharing a rare naturally occurring aldehyde functionality in this family, and a new eremophilane derivative named lithocarin A (**7**), together with two known compounds (**6** and **8**), were isolated from the deep marine sediment-derived fungus *Phomopsis lithocarpus* FS508. All of the structures for these new compounds were fully characterized and established on the basis of extensive spectroscopic interpretation and X-ray crystallographic analysis. Compound **5** exhibited cytotoxic activity against HepG-2 and A549 cell lines with IC$_{50}$ values of 16.0 and 17.6 μM, respectively.

Keywords: deep-sea derived fungus; *Phomopsis lithocarpus*; benzophenone derivatives; eremophilane derivative

1. Introduction

Fungi isolated from the deep-sea sediment are one of the most pivotal and promising source for bioactive compounds, presumably owing to the chemical diversity and biodiversity of their secondary metabolites that could be used for drug discovery and pharmacological applications [1–3]. In contrast to the terrestrial fungi, the majority of marine counterparts remain underexplored. Until now, a large number of natural products with intriguing structural skeleton and promising pharmacological effect have been discovered from marine-derived fungi, making them one of the attractive repositories for therapeutic agents and lead compounds [4–6]. Fingolimod, a synthetic derivative of immunosuppressive myriocin obtained from marine-derived *Myriococcum albomyces*, has been approved by the Food and Drug Administration and applied in patients with multiple sclerosis. Plinabulin, a synthetic analogue of the diketopiperazine halimide isolated from marine-derived *Aspergillus* sp., leading to disruption of microtubules and apoptosis, is currently examined in Phase III clinical trials for the treatment of non-small cell lung cancer. These findings have inspired natural product researchers to concentrate on the new marine compounds with pharmaceutical potential [7–12]. Fungi of the genus *Phomopsis*, widely distributed in both terrestrial and marine environments, are capable of producing structurally fascinating and architecturally diverse natural products, such as steroids [13], cytochalasins [14], pyrenocines [15], sesquiterpenes [16], diterpenes [17], terpenoids [18], oblongolides [19], which not only displayed multiple bioactivities including anti-inflammatory, antifungal, antibacterial, antiviral, antimigratory activities as well as cytotoxicity, but also provided several classes of new scaffolds that could be further modified to obtain the desired pharmaceutical effect. Accordingly, it is worthwhile to make efforts for the discovery of novel bioactive metabolites from untapped species of the genus *Phomopsis*.

Motivated by an ongoing research program focusing on biologically meaningful natural products with novel structural diversity and architectural complexity from the marine-derived fungi, numerous new compounds with cytotoxic and enzyme inhibitory activities were obtained, including lithocarpins, dichotocejpins, eutypellols and acaromycin [20–24]. Recently, a continuing investigation of *Phomopsis lithocarpus* FS508, which was isolated from the sediment sample of the India Ocean, led to obtain five new benzophenone derivatives with a rare naturally occurring aldehyde functionality named tenellones D–H (**1–5**) as well as a new eremophilane derivative lithocarin A (**7**), together with two related known compounds tenellone A [25] and AA03390 [26] (Figure 1). The structures of all the isolates were fully characterized and rationally established by comprehensive spectroscopic analyses and X-ray crystallography. All of the compounds were evaluated for their antitumor activity, and compound **5** displayed moderate inhibitory effect against human tumor cell lines HepG-2 and A549 with IC_{50} values of 16.0 and 17.6 µM, respectively.

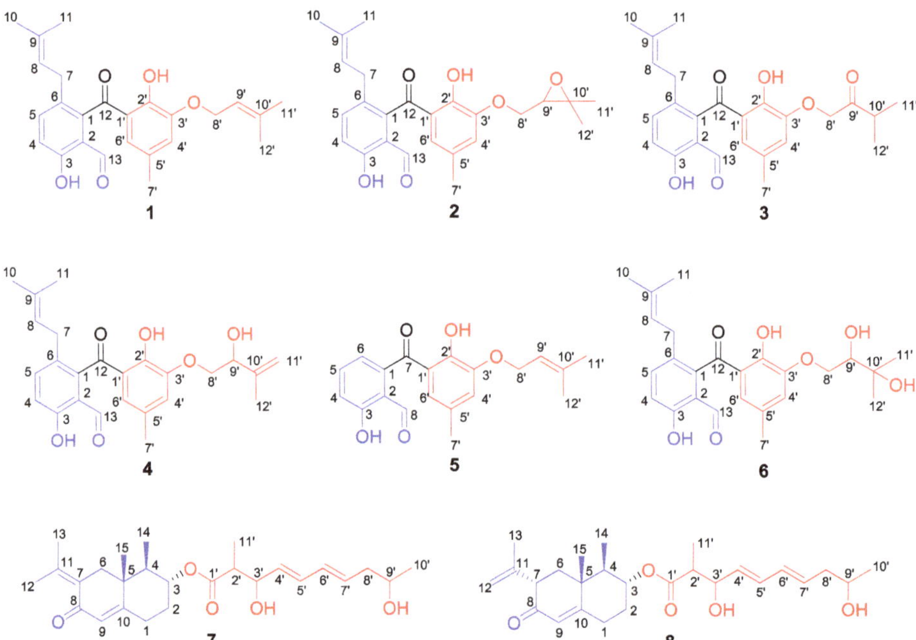

Figure 1. The structures of compounds **1–8**.

2. Results and Discussion

2.1. Structure Elucidation

Tenellone D (**1**) was obtained as yellow needles. Its molecular formula was determined as $C_{25}H_{28}O_5$ on the basis of negative HRESIMS with a deprotonated molecular ion at *m/z* 407.1867 $[M − H]^-$ (calcd. for $C_{25}H_{27}O_5$, 407.1864), indicating the existence of 12 indices of hydrogen deficiency.

The IR spectrum of **1** showed two unambiguous absorption bands at 3445 cm^{-1} and 1653 cm^{-1}, which were characteristic for hydroxy and carbonyl functionalities, respectively. With a careful inspection and analyses of its ^{13}C NMR (Table 1) and HSQC spectra (Figure S5), 25 carbon signals were successfully distinguished and ascribed to two carbonyl moieties, five methyls, two methylenes, six methines, as well as ten quaternary carbons.

Table 1. ^{1}H (600 MHz) and ^{13}C (150 MHz) NMR data of **1** and **2** in CD$_3$Cl (δ ppm, *J* in Hz).

No.	1		2	
	δ_H (*J* in Hz)	δ_C	δ_H (*J* in Hz)	δ_C
1		141.0, C		140.6, C
2		117.4, C		117.3, C
3		160.8, C		160.8, C
4	7.06, d, (8.8)	119.5, CH	7.07, d, (8.7)	119.6, CH
5	7.45, d, (8.8)	138.5, CH	7.45, d, (8.7)	138.7, CH
6		129.9, C		129.9, C
7	3.12, s	31.0, CH$_2$	3.12, s	31.0, CH$_2$
8	5.05, m	121.9, CH	5.05, m	121.7, CH
9		134.1, C		134.2, C
10	1.47, s	17.8, CH$_3$	1.47, s	17.8, CH$_3$
11	1.58, s	25.7, CH$_3$	1.59, s	25.7, CH$_3$
12		203.1, C=O		203.1, C=O
13	9.71, s	194.5, C=O	9.71, s	194.4, C=O
1′		121.0, C		121.3, C
2′		151.7, C		151.6, C
3′		148.0, C		147.7, C
4′	6.93, d, (1.9)	121.8, CH	7.03, s	123.7, CH
5′		128.6, C		129.1, C
6′	6.47, d, (1.9)	123.6, CH	6.56, s	125.1, CH
7′	2.17, s	21.1, CH$_3$	2.19, s	21.0, CH$_3$
8α′	4.62, d, (5.7)	66.3, CH$_2$	4.12, dd, (9.5, 7.8)	71.8, CH$_2$
8β′			4.43, dd, (9.5, 2.8)	
9′	5.55, m	119.4, CH	4.07, dd, (7.8, 2.8)	76.6, CH
10′		138.7, C		71.1, C
11′	1.76, s	18.4, CH$_3$	1.69, s	29.8, CH$_3$
12′	1.80, s	26.0, CH$_3$	1.71, s	28.4, CH$_3$
3-OH	11.51, s		11.50, s	
2′-OH	12.09, s		12.13, s	

The proton–proton connectivities of **1** clearly clarified the existence of three spin-coupling systems: **a** (H-4/H-5), **b** (H-7/H-8), and **c** (H-8′/H-9′) as depicted in Figure 2. On the basis of fragment **a**, a 1,2,3,6-tetrasubstituted aromatic ring was initially established attributable to the critical HMBC correlations from H-4 to C-2 and C-6 as well as H-5 to C-1 and C-3. Moreover, the HMBC correlations from H-10 and H-11 to C-9 and C-8, coupled with the fragment **b**, strongly suggested the presence of an isopentenyl fragment; and it was concluded to link at C-6 (δ_C 129.9) position in the 1,2,3,6-tetrasubstituted aromatic ring, which could be further evidenced by pivotal HMBC interactions from the methylene protons H-7 (δ_H 3.12) to C-1 and C-6. Similarly, the location of the aldehyde group C-13 (δ_C 194.5) at C-2 (δ_C 117.4) in the aromatic ring was reasonably verified by the conclusive HMBC correlations from the aldehyde proton H-13 (δ_H 9.71) to C-2 and C-3. The position of hydroxyl group was deduced to connect at C-3 mainly referred to its significant down-shifted carbon signal at δ_C 160.8, which could be further supported by the HMBC correlations from hydroxyl proton (δ_H 11.51) to C-3 and C-4. Therefore, the unit A was finally established as shown in Figure 2.

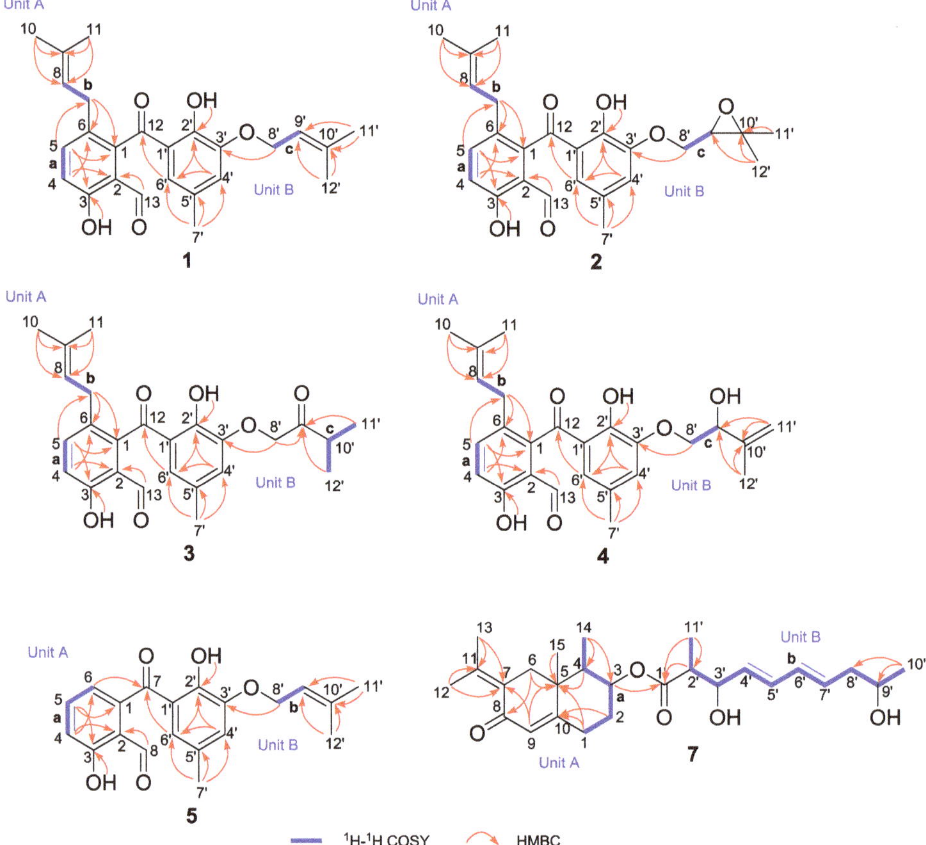

Figure 2. ^{1}H-^{1}H COSYs and key HMBCs of **1–5**, and **7**.

In unit B, the second 1′,2′,3′,5′-tetrasubstituted benzene ring with a phenolic group at C-2′ (δ_C 151.7) was readily constructed by the *meta*-coupled aromatic protons of H-4′ (δ_H 6.93) and H-6′ (δ_H 6.47), the aforementioned conclusion could be further confirmed and supported by the HMBC interactions of H-4′/C-2′, H-4′/C-6′, as well as H-6′/C-2′. In addition, the HMBC correlations from H-7′ (δ_H 2.17) to C-4′, C-5′, and C-6′ strongly suggested the location of the methyl group at C-5′ (δ_C 128.6). Furthermore, on the basis of fragment **c**, the HMBC interactions from H$_3$-11′ and H$_3$-12′ to C-9′ as well as H$_3$-11′ and H$_3$-12′ to C-10′ implied the presence of another isoprenyl moiety in unit B. This isoprenyl group could be readily assigned at C-3′ attributable to the pivotal HMBC correlation of the methylene protons H-8′ (δ_H 4.62) to C-3′ (δ_C 148.0). Therefore, the unit B was completely ascertained.

The connectivity of the units A and B was initially speculated to conjunct through a carbonyl carbon atom C-12 (δ_C 203.1) with the formation of a benzophenone architecture, which was mainly attributed to the HMBC correlation between H-6′ and C-12. However, the linkage of the unit A and the ketone group was still a critical uncertainty for the elucidation of the structure of **1** because of the lack of any conclusive long-range correlations. Fortunately, after many attempts with different solvent combinations, a single crystal of compound **1** suitable for X-ray crystallographic analysis was obtained. The further X-ray crystallographic measurement was conducted and completed on the Cu Kα, which adequately clarified the structure of **1** without ambiguity (Figure 3). In light of the aforementioned evidence, the structure of **1** was concluded as shown in Figure 2.

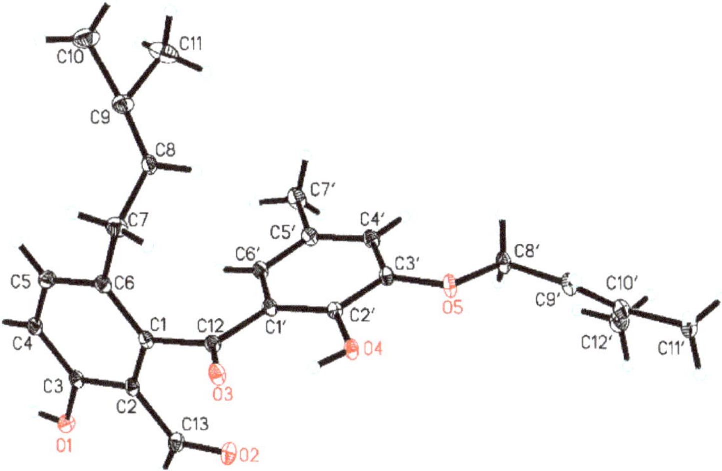

Figure 3. Perspective drawing of the X-ray structure of **1**.

Compound **2** was isolated as a yellow oil. The molecular formula was assigned as $C_{25}H_{28}O_6$ based on the negative mode HRESIMS $[M + Cl]^-$ ion at m/z 459.1583 (calcd. for 459.1580), indicating the presence of twelve degrees of unsaturation in the molecule. The ^1H and ^{13}C NMR spectra of **2** closely resembled to those of **1** except for the presence of two oxygenated alkyl carbons (δ_C 71.1, 76.6) and the absence of two olefinic carbons in **2**, suggesting that the olefinic bond should be replaced by an epoxy moiety in **2** because of the slight up-shifted proton signal (δ_H 4.07) affiliated to the oxygenated alkyl carbon, which could be also accessibly confirmed by the ^1H-^1H COSY cross peak of H-8′/H-9′ and the HMBC correlations from H-11′ (δ_C 1.69) and H-12′ (δ_C 1.71) to C-10′ (δ_C 71.1) and C-9′ (δ_C 76.6). Thus, the structure of **2** was elucidated as shown in Figure 2 and given the trivial name tenellone E.

Compound **3** was purified as a yellow oil. Its molecular formula was assigned to be the same as that for compound **2** on the basis of their HRESIMS data. The 1D NMR data (Table 2) of **3** was similar to those of **2**, indicating that they should share a very similar tenellone skeleton. After a careful inspection and close comparison, the major differences were disclosed to be the presence of a ketone functionality (δ_C 210.6) and a more high-field carbon (δ_C 37.2) in **3** instead of the epoxy moiety in **2**. As referring to the HMBC spectrum, the obvious cross peaks of H-8′ to C-9′, H-11′ to C-9′, and H-12′ to C-9′ were successfully distinguished, and it unambiguously verified the existence of a 3-methylbutan-2-one fragment in **3**. Moreover, the ^1H-^1H COSY correlations of H-10′/H-11′ and H-10′/H-12′ further accounted for the above deduction. Therefore, compound **3** was completely determined (Figure 2) and given the trivial name of tenellone F.

Tenellone G (**4**) was obtained as a yellow oil, which also shared the same molecular formula $C_{25}H_{28}O_6$ inferred from a HRESIMS with a pseudo-molecular peak at m/z 423.1825 ($[M - H]^-$, calcd. for 423.1813). From the typical NMR data between compounds **1** and **4**, it could be constructively concluded that both of them should feature a similar tenellone structure, except for the absence of a methyl group (δ_C 26.0) in **1** and the presence of a oxygenated methine (δ_C 74.0) as well as a terminal olefin (δ_C 112.2, 146.0) in **4**. These differences were attributed to the oxidative rearrangement of the isoprenyl moiety, leading to an allylic alcohol with a disubstituted double bond. This deduction could be further rationalized by the critical HMBC correlations from H-11′ (δ_H 4.92, 5.12) and H-12′ (δ_H 1.83) to C-9′ (δ_C 74.0) and C-10′ (δ_C 146.0). Thereby, the structure of **4** was ascertained as depicted in Figure 2.

Table 2. ^1H (600 MHz) and ^{13}C (150 MHz) NMR data of **3** and **4** (δ ppm, J in Hz).

No.	3 δ_H (*J* in Hz)	3 δ_C	4 δ_H (*J* in Hz)	4 δ_C
1		140.6, C		141.7, C
2		117.3, C		119.2, C
3		160.8, C		161.1, C
4	7.08, d, (8.7)	119.6, CH	7.11, d, (8.6)	119.6, CH
5	7.46, d, (8.7)	138.7, CH	7.55, d, (8.6)	139.2, CH
6		129.9, C		130.5, C
7	3.13, s	31.0, CH$_2$	3.14, d, (7.2)	31.4, CH$_2$
8	5.06, m	121.7, CH	5.08, m	122.3, CH
9		134.2, C		133.8, C
10	1.48, s	17.8, CH$_3$	1.46, s	17.6, CH$_3$
11	1.59, s	25.7, CH$_3$	1.55, s	25.7, CH$_3$
12		203.2, C=O		203.6, C=O
13	9.72, s	194.4, C=O	9.98, s	194.1, C=O
1'		121.5, C		123.2, C
2'		151.5, C		152.1, C
3'		147.0, C		148.8, C
4'	6.90, d, (1.9)	123.4, CH	7.18, d, (2.0)	123.0, CH
5'		128.8, C		129.2, C
6'	6.57, s	125.4, CH	6.69, s	124.6, CH
7'	2.16, s	21.0, CH$_3$	2.16, s	20.7, CH$_3$
8α' 8β'	4.79, s	72.9, CH$_2$	4.01, dd, (9.8, 7.5) 4.15, dd, (9.8, 3.9)	74.3, CH$_2$
9'		210.6, C=O	4.47, dd, (7.5, 3.9)	74.0, CH
10'	3.00, m	37.2, CH		146.0, C
11α' 11β'	1.20, s	18.1, CH$_3$	4.92, s 5.12, m	112.2, CH$_2$
12'	1.21, s	29.9, CH$_3$	1.83, s	19.0, CH$_3$
3-OH	11.51, s			
2'-OH	12.10, s			

Tenellone H (**5**) was obtained as a yellow powder, and its molecular formula was assigned as $C_{20}H_{20}O_5$ based on the negative mode HRESIMS with an obvious molecular ion peak at *m/z* 339.1243 ([M − H]$^-$, calcd. for 339.1238. A detailed inspection and comparison of the 1D NMR spectra between **1** and **5** clearly revealed that both of them should possess the same structural architecture, which could be further confirmed by the identical proton-proton and carbon-proton correlations (Table 3). The notable differences of two compounds were ascribed to the absence of five carbons (δ_C 17.8, 25.7, 31.0, 121.9, 134.1) in **5**, which were characteristic for an isoprenyl group, suggesting the inexistence of the isoprenyl moiety attached at C-6 in aromatic ring A for compound **5**. Moreover, the predominant correlations from H-6 to C-4, C-7, and C-2 also strengthened this conclusion. Thus, compound **5** was finally determined as a de-isoprenyl derivative of compound **1** (Figure 2).

Compound **7** was isolated as a yellow oil. Its molecular formula was deduced as $C_{26}H_{38}O_5$, which was determined on the basis of the HRESIMS peak in conjunction with ^{13}C NMR data (Table 4), requiring eight indices of hydrogen deficiency. The ^1H-^1H COSY coupled with HSQC spectra initially disclosed two spin-coupling fragments as depicted with bold blue lines in Figure 2: **a** (H-1/H-2/H-3/H-4/H-14) and **b** (H-11'/H-2'/H-3'/H-4'/H-5'/H-6'/H-7'/H-8'/H-9'/H-10'). The HMBC correlations from H-1 to C-5 and C-10 as well as H-4 to C-5, coupled with the fragment **a**, successfully established the six-membered cyclic ring with the methyl group at C-4, which could be rationally verified by the HMBC correlations of H$_3$-14 (δ_H 0.98) to C-3, C-4, and C-5.

Table 3. ^{1}H (600 MHz) and ^{13}C (150 MHz) NMR data of **5** in CD$_3$Cl (δ ppm, J in Hz).

No.	δ_H (J in Hz)	δ_C	No.	δ_H (J in Hz)	δ_C
1		142.5, C	4′	6.96, s	121.7, CH
2		117.9, C	5′		128.3, C
3		162.9, C	6′	6.69, m	124.2, CH
4	7.16, d, (8.5)	120.5, CH	7′	2.12, s	21.2, CH$_3$
5	7.59, dd, (8.5, 7.3)	136.1, CH	8′	4.62, d, (7.0)	66.3, CH$_2$
6	6.98, d, (7.3)	120.0, CH	9′	5.55, m	119.5, CH
7		201.0, C=O	10′		138.8, C
8	9.91, s	195.0, C=O	11	1.76, s	18.4, CH$_3$
1′		119.9, C	12′	1.80, s	26.0, CH$_3$
2′		152.3, C	3-OH	11.74, s	
3′		148.1, C	2′-OH	11.93, s	

Table 4. ^{1}H (500 MHz) and ^{13}C (125 MHz) NMR data of **7** in CD$_3$OD (δ ppm, J in Hz).

No.	δ_H (J in Hz)	δ_C	No.	δ_H (J in Hz)	δ_C
1α	2.34, m	30.2, CH$_2$	13	1.85, s	22.3, CH$_3$
1β	2.45, m		14	0.98, d, (6.7)	10.8, CH$_3$
2α	1.46, m	31.7, CH$_2$	15	1.03, s	17.3, CH$_3$
2β	2.16, overlapped		1′		175.2, C=O
3	4.87, td, (11.2, 4.4)	74.2, CH	2′	2.56, m	46.0, CH
4	1.67, dt, (11.2, 6.7)	46.1, CH	3′	4.24, m	74.5, CH
5		42.3, C	4′	5.59, dd, (15.0, 7.0)	131.5, CH
6α	2.16, overlapped	41.2, CH$_2$	5′	6.27, dd, (15.0, 10.4)	132.5, CH
6β	2.91, d, (13.7)		6′	6.12, dd, (15.0, 10.4)	132.5, CH
7		127.2, C	7′	5.72, dd, (15.0, 7.0)	131.5, CH
8		191.7, C=O	8′	2.26, m	42.7, CH$_2$
9	5.77, d, (1.8)	126.9, CH	9′	3.86, m	67.5, CH
10		164.9, C	10′	1.20, d (6.2)	23.1, CH$_3$
11		143.7, C	11′	1.18, d (7.2)	14.3, CH$_3$
12	2.10, s	22.8, CH$_3$			

Moreover, the HMBC showed long-range ^{1}H-^{13}C correlations from H-6 (δ_H 2.16 and 2.91) to C-5 and C-8 as well as H-9 (δ_H 5.77) to C-1, C-5, and C-7, revealing the presence of the 2-decalone substructure. The connection between methyl group at C-15 and 2-decalone moiety was deduced to be located at C-5, which was further secured by the critical HMBC interactions from H-15 to C-4, C-5, and C-6. Additionally, HMBC correlations of H-12 and H-13 to C-7 and C-11 adequately assigned the location of 2-propene at C-7. Thus, the unit A was finally determined (Figure 2). The HMBC correlations from H-2′/C-1′ and H-11′/C-1′ indicated the structural motif of 2-methyl-3,9-dihydroxydecenoic acid (C-1′-C-11′) with the aid of the fragment **b**. Consequently, the unit B was ascertained. The key cross peak of H-3 (δ_H 4.87) to C-1′ (δ_C 175.2) was observed, which unambiguously confirmed the linkage of units A and B through C-3 (δ_C 74.2). Therefore, the planar structure of **7** was determined and given the trivial name lithocarin A.

The relative configuration of **7** was clarified on the basis of the coupling constant data and NOESY spectrum (Figure 4). The observed obvious NOESY correlation of H-3/H$_3$-15 suggested that H-3 and H$_3$-15 were co-facial and arbitrarily assigned to be in β-orientation. The large $^3J_{\text{H-3,H-4}}$ (11.2 Hz) indicated the *trans*-relationship between H-3 and H-4, assigning the H-4 in α-orientation. Meanwhile, NOESY correlation from H-4 to H-6α (δ_H 2.16) implied that H-6α was α-orientation. Then, the methyl group at C-4 was β-orientation because of the NOESY correlation between H$_3$-14 to H-6β. The geometries of the double bonds (C-4′ and C-6′) could be rationally determined as 4′*E* and 6′*E* evidenced by the identical coupling constants of H-4′ and H-5′ as well as H-6′ and H-7′ (J = 15.0 Hz).

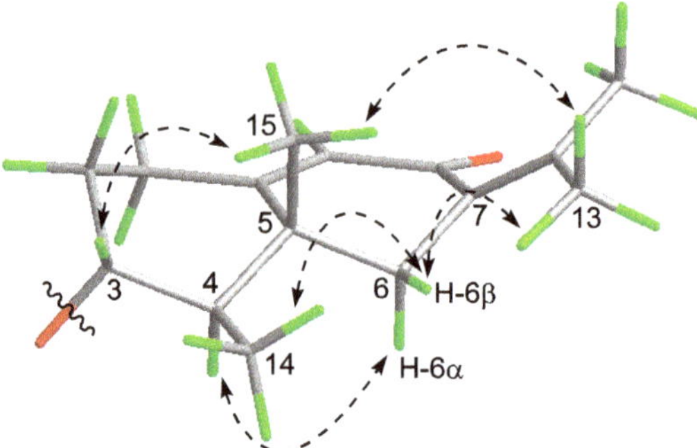

Figure 4. Key NOESY correlations of compound **7** (sesquiterpenoid moiety).

2.2. Cytotoxicity Assay

Compounds **1–8** were evaluated for their cytotoxic activities against the HepG-2, MCF-7, SF-268, and A549 tumor cell lines with cisplatin as the positive control. Compound **5** exhibited moderate inhibitory activity against HepG-2 and A549 cell lines with IC_{50} values of 16.0 and 17.6 μM, respectively; while compound **8** showed weak inhibitory effect against the tested tumor cell lines with the IC_{50} values ranging from 25.5 to 29.6 μM. The other six compounds exhibited no cytotoxic activity even at the concentration of 50 μM (Table 5).

Table 5. Cytotoxic activities of compounds **1–8**.

Compounds	IC_{50} (μM) [a]			
	HepG-2	**MCF-7**	**SF-268**	**A549**
1	>100	>100	>100	>100
2	>100	>100	>100	>100
3	>100	>100	>100	>100
4	88.6 ± 3.1	85.7 ± 7.4	67.7 ± 3.1	>100
5	16.0 ± 0.1	25.1 ± 1.1	23.0 ± 0.9	17.6 ± 0.3
6	>100	>100	>100	>100
7	90.9 ± 2.0	81.1 ± 2.8	92.5 ± 4.3	59.2 ± 2.1
8	26.2 ± 0.8	29.6 ± 4.6	28.8 ± 0.2	25.5 ± 0.4
cisplatin	2.4 ± 0.1	3.2 ± 0.1	3.3 ± 0.3	1.6 ± 0.1

[a] Values are expressed as the mean ± SD.

3. Materials and Methods

3.1. General Experimental Procedures

UV spectra were taken on a Shimadzu UV-2600 spectrophotometer (Shimadzu, Kyoto, Japan). IR data were recorded on a Shimadzu IR Affinity-1 spectrometer (Shimadzu, Kyoto, Japan). Optical rotations were measured on an Anton Paar MCP-500 spectropolarimeter (Anton Paar, Graz, Austria) at 25 °C. Circular dichroism (CD) spectra were obtained under N_2 gas on a Jasco 820 spectropolarimeter (Jasco Corporation, Kyoto, Japan). The NMR spectra were acquired using a Bruker Avance 500 MHz or 600 MHz NMR spectrometer with TMS as an internal standard (Bruker, Fällanden, Switzerland). ESIMS data were collected on an Agilent Technologies 1290-6430A Triple Quad LC/MS (Agilent Technologies, Palo Alto, CA, USA). HRESIMS were done with a Thermo MAT95XP high resolution mass spectrometer (Thermo Fisher Scientific, Bremen, Germany). Preparative HPLC separations were carried out using a YMC-pack ODS-A column (250 × 20 mm, 5 μm, 12 nm, YMC Co., Ltd, Kyoto, Japan). Semi-preparative HPLC separations were performed utilizing a YMC-pack ODS-A/AQ column (250 × 10 mm, 5 μm, 12 nm, YMC CO., Ltd, Kyoto, Japan) and a YMC-pack Cellulose-SB column (250 × 10 mm, 5 μm, 12 nm, YMC CO., Ltd, Kyoto, Japan). Column chromatography were performed with silica gel (200−300 mesh, Qingdao Marine Chemical Inc., Qingdao, China) and Sephadex LH-20 (Amersham Biosciences, Uppsala, Sweden), respectively. Thin-Layer Chromatography (TLC) was conducted with precoated glass plates GF-254 (Merck KGaA, Darmstadt, Germany).

3.2. Fungal Material

The fungus *P. lithocarpus* FS508 was isolated in 2016 from a deep-sea sediment sample collected in the Indian Ocean (111°53.335′ E, 16°50.508′ N; depth 3606 m). The sequence of amplified ITS region of the strain FS508 has been submitted to GenBank (Accession No. MG686131). A BLAST search of ITS region revealed that FS508 has 99% homology with *Phomopsis lithocarpus* CZ105B (Accession No. FJ755236). The strain is preserved at the Guangdong Provincial Key Laboratory of Microbial Culture Collection and Application, Guangdong Institute of Microbiology.

3.3. Fermentation, Extraction and Isolation

The fermentation was carried out in 3 L Erlenmeyer flasks, which contained 480 g of rice and 600 mL of 0.5% saline water. Each flask was aseptically inoculated with the seed inoculums and statically fermented for a month at 28 °C. The fermented rice substrate (10 flasks) was extracted three times with EtOAc, and the solvent was evaporated to dryness under vacuum to obtain a crude extract (99.8 g). The crude extract was completely dissolved in 80% MeOH/H_2O, which was extracted with petroleum ether four times to remove some aliphatic acids. The obtained residue was subjected to silica gel chromatography (200−300 mesh) by step gradient elution with petroleum ether/EtOAc (10:1→0:1) and followed by CH_2Cl_2/MeOH in linear gradient (8:1→0:1) to yield 14 fractions (Frs. 1−14).

Fr. 4 (0.22 g) was further fractionated by column chromatography on silica gel eluting with a *n*-hexane/EtOAc (6:1→1:1) to produce 3 fractions. Fr. 4-1 (0.06 g) was re-purified by HPLC on a semipreparative YMC-pack ODS-A/AQ column (MeOH/H_2O, 85:15, 3 mL/min) to obtain **1** (3.0 mg, t_R 21.8 min) and **5** (1.3 mg, t_R 12.1 min). Fr. 5 (0.36 g) was separated by preparative RP-HPLC system over a YMC ODS-A column (MeOH/H_2O, 90:10, 8 mL/min) to afford 3 fractions. Fr. 5-1 (0.08 g) was further fractionated on a semipreparative YMC-pack Cellulose-SB column (MeOH/H_2O, 82:18, 3 mL/min) to give 2 subfractions. Fr. 5-1-1 was purified by HPLC on a semipreparative YMC-pack ODS-A/AQ column (MeOH/H_2O, 86:14, 3 mL/min) to obtain **2** (3.2 mg, t_R 10.7 min) and **3** (1.2 mg, t_R 10.1 min). Fr. 6 (3.10 g) was fractionated by Sephadex LH-20, eluting with CH_2Cl_2/MeOH (1:1) to yield 4 fractions. Fr. 6-2 (2.20 g) was separated into 5 fractions on a silica gel column (200−300 mesh), eluting with petroleum ether/EtOAc in linear gradient (6:1→0:1). Fr. 6-2-1 (1.20 g) was re-purified by preparative RP-HPLC system over a YMC ODS-A column (MeOH/H_2O, 85:15, 8 mL/min) to produce 5 subfractions. Fr. 6-2-1-3 was further purified by HPLC on a semipreparative YMC-pack ODS-A/AQ column (MeCN/H_2O, 82:18, 3 mL/min) to obtain **4** (4.2 mg, t_R 9.8 min). Fr. 8 (15.00 g) was separated by column chromatography over C-18 reversed-phase (RP) silica gel eluting with a MeOH/H_2O gradient (30:70→100:0) to produce 13 fractions. Fr. 8-12 (1.30 g) was subjected to a silica gel column (200-300 mesh) using a stepped gradient elution of petroleum ether/EtOAc (6:1→0:1) to give 3 fractions. Fr. 8-12-1 (0.06 g) was further purified by HPLC on a semipreparative YMC-pack ODS-A/AQ column (MeOH/H_2O, 80:20, 3 mL/min) to obtain **6** (2.1 mg, t_R 12.6 min). Fr. 8-12-2 (0.36 g) was separated over Sephadex LH-20 into 4 subfractions, eluting with CH_2Cl_2/MeOH (1:1). Fr. 8-12-2-1 (0.07 g) was re-purified by HPLC on a semipreparative YMC-pack ODS-A/AQ column (MeOH/H_2O, 77:23, 3 mL/min) to afford **7** (3.9 mg, t_R 15.7 min) and **8** (10.0 mg, t_R 14.5 min).

Tenellone D (**1**): yellow needles; UV (MeOH) λ_{max} (log ε) 218.2 (4.12), 266.6 (3.60), 346.2 (3.26) nm; IR ν_{max} 3445, 2924, 1653, 1456, 1261 cm^{-1}. ^{1}H (600 MHz) and ^{13}C (150 MHz) NMR spectral data, see Table 1; negative ESIMS: *m/z* 407 [M − H]$^-$; HRESIMS: *m/z* 407.1867 [M − H]$^-$ (calcd. for $C_{25}H_{27}O_5$, 407.1864).

Tenellone E (**2**): yellow oil; $[\alpha]_D^{25}$ +10.3 (*c* 0.99, MeOH). UV (MeOH) λ_{max} (log ε) 218.6 (4.60), 266.0 (4.11), 346.2 (3.87) nm; IR ν_{max} 3502, 2926, 1653, 1456, 1265 cm^{-1}. ^{1}H (600 MHz) and ^{13}C (150 MHz) NMR spectral data, see Table 1; negative ESIMS: m/z 459 [M + Cl]$^-$; HRESIMS: m/z 459.1583 [M + Cl]$^-$ (calcd. for C$_{25}$H$_{28}$O$_6$Cl, 459.1580).

Tenellone F (**3**): yellow oil; UV (MeOH) λ_{max} (log ε) 216.0 (4.40), 267.4 (3.92), 345.4 (3.63) nm; IR ν_{max} 3377, 2924, 1651, 1456, 1267 cm^{-1}. ^{1}H (600 MHz) and ^{13}C (150 MHz) NMR spectral data, see Table 2; negative ESIMS: m/z 423 [M − H]$^-$; HRESIMS: m/z 423.1808 [M − H]$^-$ (calcd. for C$_{25}$H$_{27}$O$_6$, 423.1813).

Tenellone G (**4**): yellow oil; $[\alpha]_D^{25}$ +10.0 (*c* 1.02, MeOH). UV (MeOH) λ_{max} (log ε) 218.6 (4.61), 266.0 (4.12), 346.0 (3.88) nm; IR ν_{max} 3356, 2926, 1647, 1450, 1273 cm^{-1}. ^{1}H (600 MHz) and ^{13}C (150 MHz) NMR spectral data, see Table 2; negative ESIMS: m/z 423 [M − H]$^-$; HRESIMS: m/z 423.1825 [M − H]$^-$ (calcd. for C$_{25}$H$_{27}$O$_6$, 423.1813).

Tenellone H (**5**): yellow powder; UV (MeOH) λ_{max} (log ε) 213.0 (4.58), 265.8 (4.06), 335.4 (3.75) nm; IR ν_{max} 3377, 2931, 1651, 1474, 1022 cm^{-1}. ^{1}H (600 MHz) and ^{13}C (150 MHz) NMR spectral data, see Table 3; negative ESIMS: m/z 339 [M − H]$^-$; HRESIMS: m/z 339.1243 [M − H]$^-$ (calcd. for C$_{20}$H$_{20}$O$_5$, 339.1238).

Lithocarin A (**7**): yellow oil; $[\alpha]_D^{25}$ −16.3 (*c* 0.98, MeOH). UV (MeOH) λ_{max} (log ε) 204.0 (4.37), 235.2 (4.50), 276.0 (4.12) nm; CD (0.20 mg/mL, MeOH) λ_{max} ($\Delta\varepsilon$) 215 (−11.40), 241 (+35.00), 282 (−16.67), 326 (−1.73) nm; IR ν_{max} 3366, 2939, 1717, 1655, 1373 cm^{-1}. ^{1}H (500 MHz) and ^{13}C (125 MHz) NMR spectral data, see Table 4; positive ESIMS: m/z 453 [M + Na]$^+$; HRESIMS: m/z 453.2609 [M + Na]$^+$ (calcd. for C$_{26}$H$_{38}$O$_5$, 453.2611).

3.4. X-Ray Analysis of Tenellone D (1)

The single-crystal X-ray diffraction data of compound **1** was collected at 100 K on Rigaku Oxford Diffraction Supernova Dual Source, Cu at Zero equipped with an AtlasS2 CCD using Cu Kα radiation. Data reduction was carried out with the diffractometer's software. The structures were solved by direct methods using Olex2 software, and the non-hydrogen atoms were located from the trial structure and then refined anisotropically with SHELXL-2014 using a full-matrix least squares procedure based on F^2. The weighted R factor, wR and goodness-of-fit S values were obtained based on F^2. The hydrogen atom positions were fixed geometrically at the calculated distances and allowed to ride on their parent atoms. Crystallographic data for the structure of tenellone D (**1**) reported in this paper has been deposited in the Cambridge Crystallographic Data Centre. (Deposition number: CCDC 1852781). Copies of these data can be obtained free of charge via www.ccdc.cam.au.ck/conts/retrieving.html.

Crystal data for compound **1**: C$_{25}$H$_{28}$O$_5$, M = 408.47, monoclinic, size 0.13 × 0.12 × 0.11 mm^3, space group P2$_1$/c; a = 9.9962 (3) Å, b = 21.9296 (6) Å, c = 10.3159 (3) Å, α = 90.00°, β = 109.354(4), γ = 90.00°, V = 2133.57 (12) Å^3, T = 100.00 K, Z = 4, $\rho_{calcd.}$ = 1.272 g/cm^3, $F(000)$ = 872.0, 8355 reflections in $-12 \leq h \leq 6$, $-26 \leq k \leq 26$, $-12 \leq l \leq 12$, measured in the range $8.064° \leq 2\theta \leq 147.114°$, GOOF = 1.036, Final R indices I > 2σ(I): R_1 = 0.0494, wR$_2$ = 0.1245, Final R indices (all data): R_1 = 0.0579, wR$_2$ = 0.1328, largest difference peak and hole = 0.27 and −0.24 e. Å^{-3}.

3.5. Cytotoxicity Assay

The cytotoxic activities of compounds (**1–8**) were evaluated against four human tumor cell lines HepG-2, MCF-7, SF-268, and A549 with cisplatin as the positive control. Assays were performed by the Sulforhodamine (SRB) method [27].

4. Conclusions

In summary, five new benzophenone derivatives and a new eremophilane derivative, along with two known compounds, were isolated from the marine-derived fungus *P. lithocarpus* FS508. The chemical structures of eight compounds were elucidated by means of NMR analyses and single-crystal X-ray diffraction. The antitumor activities of compounds **1–8** were evaluated wherein compound **5** exhibited moderate growth inhibition against HepG-2 and A549 cell lines with IC_{50} values of 16.0 and 17.6 µM, respectively; while compound **8** displayed weak inhibitory effect against four human tumor cell lines with the IC_{50} values ranging from 25.5 to 29.6 µM. However, compounds **1–4**, **6** and **7** were inactive against these tumor cell lines even at 50 µM. Comparing with compound **5**, compounds **1–4** and **6**, possessing an isoprenyl group in the unit A, failed to show a cytotoxic effect, suggesting that the isoprenyl group might impede the cytotoxic activity. It's worth to note that compound **8** showed better cytotoxicity than **7**, which indicated that the position of the olefinic bond should play a vital influence on their biological activity.

Supplementary Materials: The following are available online at http://www.mdpi.com/1660-3397/16/9/329/s1. Figures S1–S4: HRESIMS, IR, UV, CD and 1D and 2D NMR spectra of compounds **1–5**, and **7**; Figures S55–S58: 1D NMR spectra of compounds **6** and **8**.

Author Contributions: J.-L.X., H.-X.L. and H.-B.T. elucidated structures and wrote the paper; J.-L.X., H.-X.L., H.G. and Z.-L.H. performed the fermentation, extraction and isolation experiments; Y.-C.C., H.-H.L., L.-Q.X. and S.-N.L. performed biological evaluations experiments; W.-M.Z. and X.-X.G. designed and coordinated the study and reviewed the manuscript. All authors have read and approved the manuscript.

Funding: This work was supported financially by the Science and Technology Program of Guangzhou, China (201607020018), the Team Project of the Natural Science Foundation of Guangdong Province (2016A030312014), the National Natural Science Foundation of China (31272087), and the Guangdong Provincial Innovative Development of Marine Economy Regional Demonstration Projects (GD2012-D01-002).

Acknowledgments: We sincerely thank Can Li of Central Laboratory of Southern Medical University for NMR measurement.

Conflicts of Interest: The author declare no conflicts.

References

1. Li, X.D.; Li, X.M.; Li, X.; Xu, G.M.; Liu, Y.; Wang, B.G. Aspewentins D–H, 20-nor-isopimarane derivatives from the deep sea sediment-derived fungus *Aspergillus wentii* SD-310. *J. Nat. Prod.* **2016**, *79*, 1347–1353. [CrossRef] [PubMed]

2. Wang, C.; Guo, L.; Hao, J.J.; Wang, L.P.; Zhu, W.M. α-glucosidase inhibitors from the marine-derived fungus *Aspergillus flavipes* HN4-13. *J. Nat. Prod.* **2016**, *79*, 2977–2981. [CrossRef] [PubMed]

3. Deshmukh, S.K.; Prakash, V.; Ranjan, N. Marine fungi: A source of potential anticancer compounds. *Front. Microbiol.* **2018**, *8*, 2536. [CrossRef] [PubMed]

4. Liao, L.J.; Bae, S.Y.; Won, T.H.; You, M.J.; Kim, S.H.; Oh, D.C.; Lee, S.K.; Oh, K.B.; Shin, J. Asperphenins A and B, lipopeptidyl benzophenones from a marine-derived *Aspergillus* sp. fungus. *Org. Lett.* **2017**, *19*, 2066–2069. [CrossRef] [PubMed]

5. Soldatou, S.; Baker, B.J. Cold-water marine natural products, 2006 to 2016. *Nat. Prod. Rep.* **2017**, *34*, 585–626. [CrossRef] [PubMed]

6. Dyshlovoy, S.A.; Honecker, F. Marine compounds and cancer: 2017 updates. *Mar. Drugs* **2018**, *16*, 41. [CrossRef] [PubMed]

7. Wang, Y.T.; Xue, Y.R.; Liu, C.H. A brief review of bioactive metabolites derived from deep-sea fungi. *Mar. Drugs* **2015**, *13*, 4594–4616. [CrossRef] [PubMed]

8. Blunt, J.W.; Carroll, A.R.; Copp, B.R.; Davis, R.A.; Keyzers, R.A.; Prinsep, M.R. Marine natural products. *Nat. Prod. Rep.* **2018**, *35*, 8–53. [CrossRef] [PubMed]

9. Bao, J.; Zhai, H.J.; Zhu, K.K.; Yu, J.H.; Zhang, Y.Y.; Wang, Y.Y.; Jiang, C.S.; Zhang, X.Y.; Zhang, Y.; Zhang, H. Bioactive pyridone alkaloids from a deep-sea-derived fungus *Arthrinium* sp. UJNMF0008. *Mar. Drugs* **2018**, *16*, 174. [CrossRef] [PubMed]

10. Meng, J.J.; Cheng, W.; Heydari, H.; Wang, B.; Zhu, K.; Konuklugil, B.; Lin, W.H. Sorbicillinoid-based metabolites from a sponge-derived fungus *Trichoderma saturnisporum*. *Mar. Drugs* **2018**, *16*, 226. [CrossRef] [PubMed]

11. Daletos, G.; Ebrahim, W.; Ancheeva, E.; El-Neketi, M.; Song, W.G.; Lin, W.H.; Proksch, P. Natural products from deep-sea-derived fungi a new source of novel bioactive compounds? *Curr. Med. Chem.* **2018**, *25*, 186–207. [CrossRef] [PubMed]

12. Wang, Y.X.; Zhang, H.; Gigant, B.; Yu, Y.M.; Wu, Y.P.; Chen, X.Z.; Lai, Q.H.; Yang, Z.Y.; Chen, Q.; Yang, J.L. Structures of a diverse set of colchicine binding site inhibitors in complex with tubulin provide a rationale for drug discovery. *FEBS J.* **2016**, *283*, 102–111. [CrossRef] [PubMed]

13. Hu, Z.X.; Wu, Y.; Xie, S.S.; Sun, W.G.; Guo, Y.; Li, X.N.; Liu, J.J.; Li, H.; Wang, J.P.; Luo, Z.W.; et al. Phomopsterones A and B, two functionalized ergostane-type steroids from the endophytic fungus *Phomopsis* sp. TJ507A. *Org. Lett.* **2017**, *19*, 258–261. [CrossRef] [PubMed]

14. Shang, Z.; Raju, R.; Salim, A.A.; Khalil, Z.G.; Capon, R.J. Cytochalasins from an Australian marine sediment-derived *Phomopsis* sp. (CMB-M0042F): Acid-mediated intramolecular cycloadditions enhance chemical diversity. *J. Org. Chem.* **2017**, *82*, 9704–9709. [CrossRef] [PubMed]

15. Hussain, H.; Krohn, K.; Ahmed, I.; Draeger, S.; Schulz, B.; Di Pietro, S.; Pescitelli, G. Phomopsinones A–D: Four new pyrenocines from endophytic fungus *Phomopsis* sp. *Eur. J. Org. Chem.* **2012**, *2012*, 1783–1789. [CrossRef]

16. Xie, S.S.; Wu, Y.; Qiao, Y.B.; Guo, Y.; Wang, J.P.; Hu, Z.X.; Zhang, Q.; Li, X.N.; Huang, J.F.; Zhou, Q.; et al. Protoilludane, illudalane, and botryane sesquiterpenoids from the endophytic fungus *Phomopsis* sp. TJ507A. *J. Nat. Prod.* **2018**, *81*, 1311–1320. [CrossRef] [PubMed]

17. Wei, W.; Gao, J.; Shen, Y.; Chu, Y.L.; Xu, Q.; Tan, R.X. Immunosuppressive diterpenes from *Phomopsis* sp. S12. *Eur. J. Org. Chem.* **2014**, *2014*, 5728–5734. [CrossRef]

18. Li, L.Y.; Sattler, I.; Deng, Z.W.; Groth, I.; Walther, G.; Menzel, K.D.; Peschel, G.; Grabley, S.; Lin, W.H. A-seco-oleane-type triterpenes from *Phomopsis* sp. (strain HKI0458) isolated from the mangrove plant *Hibiscus tiliaceus*. *Phytochemistry* **2008**, *69*, 511–517. [CrossRef] [PubMed]

19. Bunyapaiboonsri, T.; Yoiprommarat, S.; Srikitikulchai, P.; Srichomthong, K.; Lumyong, S. Oblongolides from the endophytic fungus *Phomopsis* sp. BCC9789. *J. Nat. Prod.* **2010**, *73*, 55–59. [CrossRef] [PubMed]

20. Gao, X.W.; Liu, H.X.; Sun, Z.H.; Chen, Y.C.; Tan, Y.Z.; Zhang, W.M. Secondary metabolites from the deep-sea derived fungus *Acaromyces ingoldii* FS121. *Molecules* **2016**, *21*, 371. [CrossRef]

21. Fan, Z.; Sun, Z.H.; Liu, Z.; Chen, Y.C.; Liu, H.X.; Li, H.H.; Zhang, W.M. Dichotocejpins A–C: New diketopiperazines from a deep-sea-derived fungus *Dichotomomyces cejpii* FS110. *Mar. Drugs* **2016**, *14*, 164. [CrossRef] [PubMed]

22. Liu, H.X.; Zhang, L.; Chen, Y.C.; Li, S.N.; Tan, G.H.; Sun, Z.H.; Pan, Q.L.; Ye, W.; Li, H.H.; Zhang, W.M. Cytotoxic pimarane-type diterpenes from the marine sediment-derived fungus *Eutypella* sp. FS46. *Nat. Prod. Res.* **2017**, *31*, 404–410. [CrossRef] [PubMed]

23. Liu, H.X.; Zhang, L.; Chen, Y.C.; Sun, Z.H.; Pan, Q.L.; Li, H.H.; Zhang, W.M. Monoterpenes and sesquiterpenes from the marine sediment-derived fungus *Eutypella scoparia* FS46. *J. Asian Nat. Prod. Res.* **2017**, *19*, 145–151. [CrossRef] [PubMed]

24. Xu, J.L.; Tan, H.B.; Chen, Y.C.; Li, S.N.; Huang, Z.L.; Guo, H.; Li, H.H.; Gao, X.X.; Liu, H.X.; Zhang, W.M. Lithocarpins A–D: Four tenellone-macrolide conjugated [4 + 2] hetero-adducts from the deep-sea derived fungus *Phomopsis lithocarpus* FS508. *Org. Chem. Front.* **2018**, *5*, 1792–1797. [CrossRef]

25. Zhang, C.W.; Ondeyka, J.G.; Herath, K.B.; Guan, Z.Q.; Collado, J.; Platas, G.; Pelaez, F.; Leavitt, P.S.; Gurnett, A.; Nare, B.; et al. Tenellones A and B from a *Diaporthe* sp. Two highly substituted benzophenone inhibitors of parasite cGMP-dependent protein kinase activity. *J. Nat. Prod.* **2005**, *68*, 611–613. [CrossRef] [PubMed]

26. Jiang, W.D.; Jiang, Z.D.; Gallagher, R.T. Pine root (Deuteromycetes) fungal metabolites and analogs and derivatives thereof for anticancer agents. USA 5932613 A, 1999.

27. Skehan, P.; Storeng, R.; Scudiero, D.; Monks, A.; McMahon, J.; Vistica, D.; Warren, J.T.; Bokesch, H.; Kenney, S.; Boyd, M.R. New colorimetric cytotoxicity assay for anticancer-drug screening. *J. Natl. Cancer Inst.* **1990**, *82*, 1107–1112. [CrossRef] [PubMed]

 *marine drugs*

Article

Sorbicillasins A–B and Scirpyrone K from a Deep-Sea-Derived Fungus, *Phialocephala* sp. FL30r

Zhenzhen Zhang [1], Xueqian He [1], Qian Che [1], Guojian Zhang [1,2,*], Tianjiao Zhu [1], Qianqun Gu [1] and Dehai Li [1,2,*]

[1] Key Laboratory of Marine Drugs, Chinese Ministry of Education, School of Medicine and Pharmacy, Ocean University of China, Qingdao 266003, China; 15192768701@163.com (Z.Z.); h19491001@163.com (X.H.); cheqian064@ouc.edu.cn (Q.C.); zhutj@ouc.edu.cn (T.Z.); guqianq@ouc.edu.cn (Q.G.)

[2] Laboratory for Marine Drugs and Bioproducts of Qingdao National Laboratory for Marine Science and Technology, Qingdao 266237, China

* Correspondence: zhangguojian@ouc.edu.cn (G.Z.); dehaili@ouc.edu.cn (D.L.); Tel.: +86-532-82031619 (D.L.)

Received: 6 June 2018; Accepted: 18 July 2018; Published: 23 July 2018

Abstract: Two new nitrogen-containing sorbicillinoids named sorbicillasins A and B (**1** and **2**) and a new 3,4,6-trisubstituted α-pyrone derivative, scirpyrone K (**3**), together with two known biosynthetically related polyketides (**4–5**), were isolated from the deep-sea-derived fungus *Phialocephala* sp. FL30r by using the OSMAC (one strain-many compounds) method. The structures of **1–3**, including absolute configurations, were deduced based on MS, NMR, and time-dependent density functional theory (TD-DFT) calculations of specific ECD (electronic circular dichroism) spectra. Compounds **1** and **2** possessed a novel hexahydropyrimido[2,1-*a*] isoindole moiety, and compound **3** exhibited weak radical scavenging activity against DPPH (2,2-diphenyl-1-picrylhydrazyl) with an IC$_{50}$ value of 27.9 μM.

Keywords: deep-sea derived fungus; *Phialocephala* sp.; nitrogen-containing sorbicillinoids; radical scavenging activity

1. Introduction

Filamentous fungi are known as prolific microbial factories for the production of a wide range of metabolites having extensive biological activity [1]. However, previous genomic analysis of fungi revealed a large number of biosynthetic genes that were unexpressed under common laboratory culture conditions, which offers a great opportunity for natural product discovery research [2]. To induce the expression of silent biogenetic clusters and increase the chemical diversity of the secondary metabolome, the approach "one strain-many compounds (OSMAC)" has been widely and successfully practiced by altering media constituents and manipulating culture conditions [3–5].

During our ongoing search for bioactive secondary metabolites from deep-sea-derived fungi [6], the fungal strain *Phialocephala* sp. FL30r was found to be an extensive producer of diverse polyketides, including monomeric sorbicillinol derivatives, bisorbicillinoids, and trisorbicillinoids [7–10]. Based upon the biosynthetic capability of this strain, the OSMAC approach was employed to further enhance the structural diversity of secondary metabolites. When the fungal strain was cultured on a mannitol-based medium, the HPLC-UV profile (Figure S1) of the fungal extract differed from those generated previously from culture in a potato-based medium [7–10]. Further chemical assessment of the organic extract led to the isolation of two new nitrogen-containing sorbicillinoids named sorbicillasins A and B (**1** and **2**) and a new 3,4,6-trisubstituted α-pyrone derivative, scirpyrone K (**3**), together with two known biosynthetically related polyketides (**4–5**) [11,12]. Among them, compounds **1** and **2** are sorbicillin-asparagine hybrids possessing a unique hexahydropyrimido[2,1-*a*] isoindole tricyclic skeleton. The radical scavenging activities of the new compounds against DPPH (2,2-diphenyl-1-picrylhydrazyl) were tested, and **3** showed weak activity with an IC_{50} value of 27.9 μM. Herein, we report the details of the isolation, structure elucidation, and biological activities of compounds **1**–**5**.

2. Results and Discussion

The fungus *Phialocephala* sp. FL30r was cultured in mannitol-based medium (45.0 L) with agitation. The EtOAc extract (15.0 g) of fermentation was fractionated by silica gel column chromatography, Sephadex LH-20 column chromatography, medium-pressure preparation liquid chromatography (MPLC; ODS), and semi-preparative HPLC to afford compounds **1** (4.5 mg), **2** (2.7 mg), **3** (7.0 mg), **4** (4.4 mg), and **5** (2.0 mg) (Figure 1).

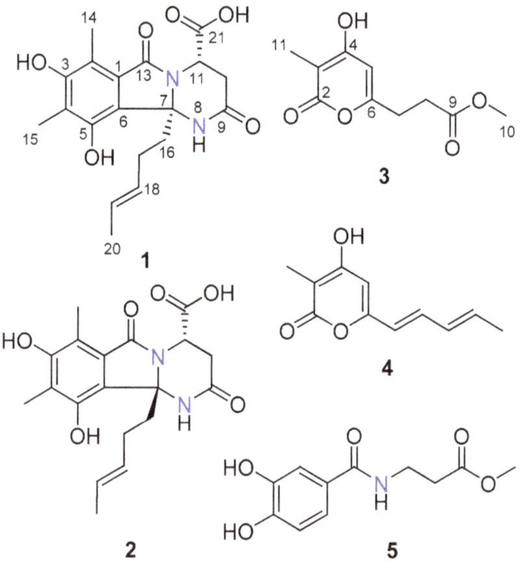

Figure 1. Structures of compounds **1**–**5**.

Sorbicillasins A and B (**1** and **2**) were both obtained as yellow oils with the molecular formula $C_{19}H_{22}N_2O_6$ according to the protonated peak at m/z 375.1550 (Figure S9, calcd. for $C_{19}H_{23}N_2O_6$, 375.1151) and the sodinated peak at m/z 397.1376 (Figure S17, calcd. for $C_{19}H_{22}N_2O_6Na$, 397.1370) as analyzed by HRESIMS, respectively. The 1D NMR data (Table 1) of **1** and **2** were highly similar. Each set of data suggested the presence of 19 carbons, assigned as 3 methyls, 3 methylenes, 3 methines with 2 vinyl moieties, and 10 non-protonated carbons, including 3 carbonyls. Among the three carbonyls, one was proven to be a carboxylic group based on the exchangeable proton signal at δ_H 13.12, together with the IR absorptions at 3524 and 1670 cm^{-1}. The planar structures of **1** and **2** were determined to be the same by interpretation of 1D and 2D NMR spectroscopic data (Figure 2, Table 1, Figures S2–S7, and Figures S11–S15). The methylated olefinic hydrocarbon chain (from C-16 to C-20) was established by the sequential COSY correlations of H_2-16/H_2-17/H-18/H-19/H_3-20. The presence of a fully substituted benzene ring was indicated by the aromatic non-protonated carbon signals (C-1/C-6) in the ^{13}C-NMR spectrum. Consistent with this analysis, diagnostic HMBC correlations were observed from H_3-14 to C-1, C-2, and C-3; from H_3-15 to C-3, C-4, and C-5; from 3-OH to C-2, C-3, and C-4; and from 5-OH to C-6. The presence of a tetrahydro-pyrimidinone ring was postulated based on the COSY correlation (H-10/H-11) and the HMBC correlations from 8-NH to C-7, C-9, and C-10;

from H$_2$-10 to C-9 and C-11; and from H-11 to C-7, C-9, and C-10. The HMBC correlations from 8-NH and H$_2$-16 to C-6 confirmed the connection of the tetrahydro-pyrimidinone structure and the benzene ring. The HMBC correlation from H$_2$-16 to C-7 positioned the olefinic chain at C-7. Based on the key HMBC correlations from H-11 to C-13 and C-21, together with the chemical shift of C-11 (δ_C 48.2) and C-13 (δ_C 169.0), two carbonyls (C-13 and C-21) were connected to N-12 and C-11, respectively. Finally, when accounting for the molecular formula and the degree of unsaturation, C-1 was linked to C-13, and one hydroxyl group was attached to C-21, thus completing the planar structure of **1**.

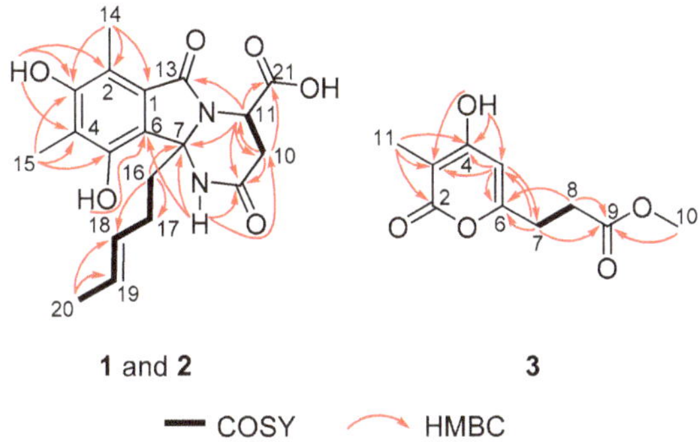

Figure 2. Key COSY and HMBC correlations of compounds **1–3**.

The relative configuration of **1** was deduced based on the NOESY correlations (Figures 3 and S8). The *E* geometries of double bonds in the olefinic chain were deduced by the correlations between H-17 and H-19 and between H-18 and H-20. The NOEs of H-10a/H-11 and H-10b/H-16 indicated that the carboxylic group and methylated olefinic hydrocarbon chain were to the same face of the pyrimidinone ring. Thus, the relative configuration of **1** was suggested to be 7*R**, 11*S**. The absolute configuration of **1** was determined by comparing the experimental ECD curve with the one calculated from the truncated model (7*R*,11*S*)-**1a** using time-dependent density functional theory (TD-DFT). The DFT re-optimization of the initial MMFF (Merck molecular force field) minima was performed at the B3LYP/6-31+g(d) level with a polarizable continuum model (PCM) solvent for MeOH. The strong agreement between the calculated ECD spectra of (7*R*,11*S*)-**1a** with experimental results suggested the absolute configuration of **1** as 7*R*, 11*S* (Figures 4 and S27).

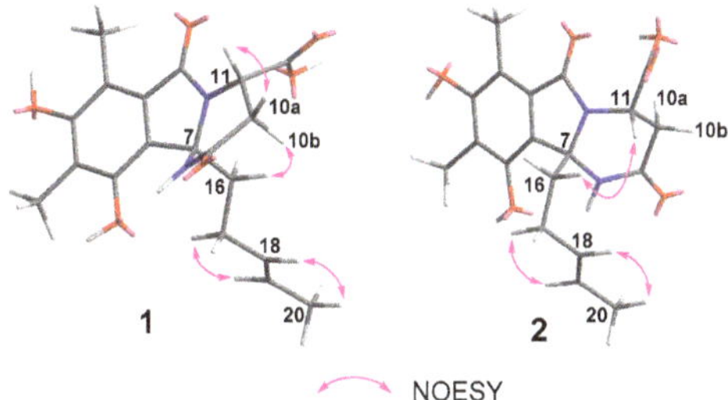

Figure 3. Key NOESY correlations of compounds **1** and **2**.

The slight discrepancies of **1** and **2** in the NMR data suggested they might be isomers. Further NOESY correlation (Figures 3 and S16) of H-11/H$_2$-16 indicated that the relative configuration of **2** was 7*S**, 11*S**. The absolute configuration of **2** was then further determined as 7*S*, 11*S* by the agreement between the calculated ECD spectra of **2** and the experimental results according to the truncated model (7*S*,11*S*)-**1b** (Figures 5 and S28).

Table 1. ^{1}H (500 MHz) and ^{13}C (125 MHz) NMR data of compounds **1** and **2** (DMSO-d_6, δ ppm).

No.	**1**		**2**	
	δ_C	δ_H, **Mult. (*J* in Hz)**	δ_C	δ_H, **Mult. (*J* in Hz)**
1	125.6		125.6	
2	115.6		115.8	
3	155.7		155.8	
4	118.9		119.2	
5	148.4		148.4	
6	124.8		130.7	
7	75.2		75.2	
9	167.1		168.0	
10	31.0	2.48 dd (17.1, 6.0), H-10a 2.82 dd (17.1, 6.0), H-10b	34.0	2.35 m, H-10a 2.61 m, H-10b
11	48.2	4.94 t (7.55)	50.2	4.31 s
13	169.0		169.5	
14	10.1	2.38 s	10.2	2.33 s
15	10.7	2.11 s	10.7	2.10 s
16	36.9	2.41 dt (11.8, 4.0) 1.90 dt (11.8, 4.0)	35.9	2.45 m 2.13 m
17	27.1	1.80 dt (11.8, 5.8) 1.12 dt (11.8, 5.8)	27.3	1.66 m 1.32 m
18	130.4	5.19 m	130.5	5.24 m
19	124.8	5.17m	125.1	5.19 m
20	18.2	1.48 d (4.2)	18.3	1.48 d (5.7)
21	172.6		171.4	
3-OH		8.62 s		8.60 s
5-OH		8.77 s		8.76 s
8-NH		8.04 s		8.03 s
21-OH		13.12 brs		12.74 brs

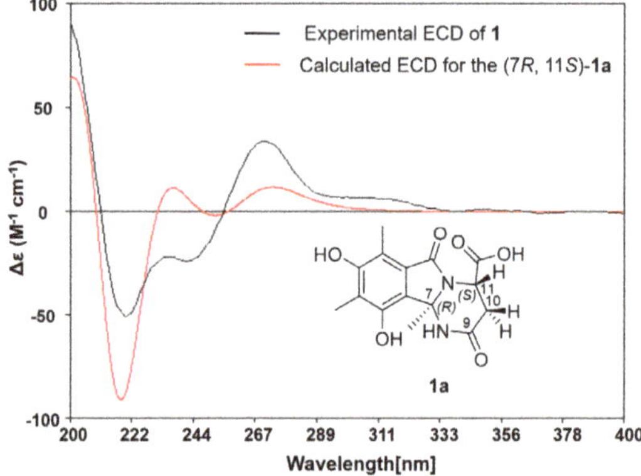

Figure 4. Experimental ECD spectrum of **1** (black curve) and that calculated from the truncated model **1a** (red curve) (0.30 eV).

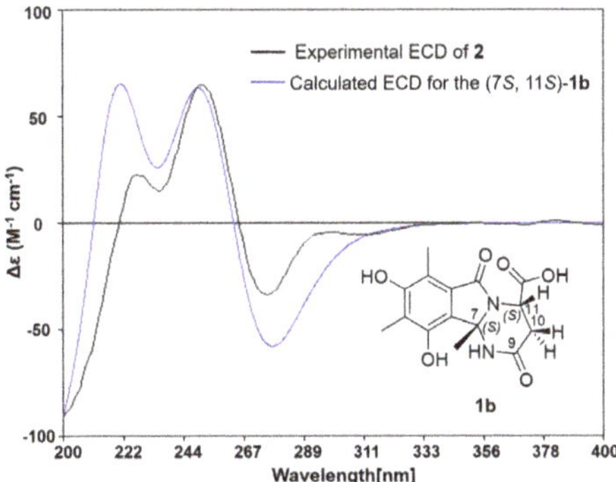

Figure 5. Experimental ECD spectrum of **2** (black curve) and that calculated from the truncated model **1b** (blue curve) (0.35 eV).

Compound **3** was obtained as a white amorphous powder, and the molecular formula was determined to be $C_{10}H_{12}O_5$ by HRESIMS peaks at m/z 213.0764 (Figure S25, calcd. for $C_{10}H_{13}O_5$, 213.0757). The 1D NMR data (Table 2) of **3** suggested the presence of two methyls, including one methoxy (δ_C 52.0 and δ_H 3.58), two methylenes, one methine, and five non-protonated carbons. Comparison of the ^{1}H and ^{13}C NMR spectra (Figures S19–S21) of **3** with those of scirpyrone H revealed the presence of an extra methyl group (δ_C 8.8 and δ_H 1.72) and the replacement of the 4-methoxyl group by a 4-hydroxyl group in **3** [13]. Further 2D NMR (Figures S22–S24) data and key HMBC correlations from H$_3$-11 to C-2, C-3, and C-4, from H-5 to C-3, C-4, and C-6, and from 4-OH to C-3, C-4, and C-5 supported the locations of the 4-hydroxyl group and the 11-methyl group.

Table 2. ^{1}H (500 MHz) and ^{13}C (125 MHz) NMR data of compound **3** (DMSO-d_6, δ ppm).

No.	δ_C	δ_H, Mult. (*J* in Hz)
2	165.3	
3	97.3	
4	165.2	
5	99.9	5.98 s
6	161.2	
7	28.4	2.67 t (6.95)
8	30.2	2.59 t (6.95)
9	172.4	
10	52.0	3.58 s
11	8.8	1.72 s
4-OH		11.15 brs

^{a} Recorded in DMSO.

Compounds **4** and **5** were identified as trichopyrone [11] and peniginseng A [12] based on the comparison of their spectroscopic data (NMR and MS) with those reported in the literature.

The cytotoxicity against K562 and MGC-803 cell lines [14,15] and the radical scavenging activity [16] against DPPH of the new compounds **1–3** were evaluated. All of them were non-cytotoxic. Compound **3** exhibited weak activity against DPPH with an IC$_{50}$ value of 27.9 μM (ascorbic acid was used as a positive control with an IC$_{50}$ value of 14.2 μM), whereas compounds **1** and **2** were not active (IC$_{50}$ > 500 μM). According to the literature [11,12], the known compound **4** showed weak radical scavenging activity, but the radical scavenging activity of **5** has not been reported.

Sorbicillinoids belong to a large family of polyketides with highly diverse carbon skeletons and bioactivities [17]. Since first reported in 1948, about 90 sorbicillinoids have been isolated from terrestrial- and marine-derived fungi [17]. Among them, the nitrogen-containing analogues are rare, with only eight related cases reported, including sorbicillactones A and B [18], sorbicillinoid urea [19], and sorbicillamines A–E [20]. The sources of nitrogen atoms in the reported nitrogenous sorbicillinoids were deduced to be L-alanine, urea, and an aminotransferase enzyme [18–20]. In this report, sorbicillasins A and B (**1** and **2**) were probably formed by adding a whole molecule of L-asparagine to 2′,3′-dihydrosorbicillin [21] (Figure 6) via sequential intermolecular/intramolecular nucleophilic reactions. The hexahydropyrimido[2,1-*a*] isoindole ring system in compounds **1** and **2**, which compose a 6/5/6 tricyclic ring system, have not been found in nature, with only related synthetic structures reported [22–24]. The above result shows that the OSMAC approach is a useful method to discover structurally diversified metabolites from a deep-sea-derived fungal strain.

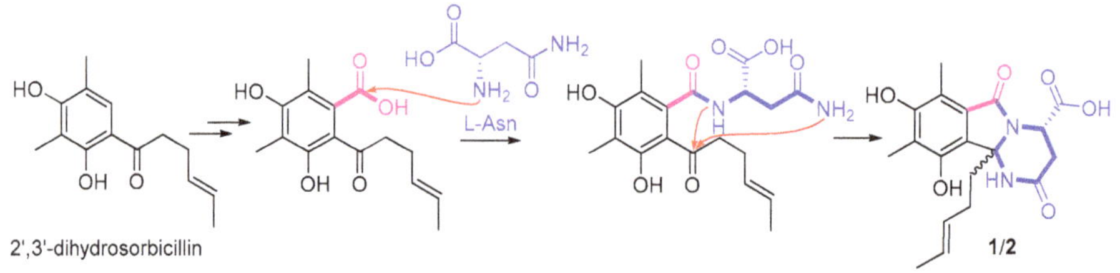

Figure 6. Possible biosynthetic pathway of compounds **1** and **2**.

3. Materials and Methods

3.1. General Experimental Procedures

UV spectra were recorded on a Beckman DU 640 spectrophotometer (Beckman Coulter Inc., Brea, CA, USA). Specific rotations were measured with a JASCO P-1020 digital polarimeter (JASCO Corporation, Tokyo, Japan). ESIMS were obtained on a Thermo Scientific LTQ Orbitrap XL mass spectrometer (Thermo Fisher Scientific, Waltham, MA, USA) or a Micromass Q-TOF ULTIMA GLOBAL GAA076 LC Mass spectrometer (Wasters Corporation, Milford, MA, USA). CD spectra were measured on a JASCO J-815 spectropolarimeter (JASCO Corporation, Tokyo, Japan). NMR spectra were recorded with an Agilent 500 MHz DD2 spectrometer (Agilent Technologies Inc., Santa Clara, CA, USA) using TMS as an internal standard, and chemical shifts were recorded as δ-values. Semi-preparative HPLC (Hitachi, Tokyo, Japan) was performed on an ODS column (YMC-Pack ODS-A, 10 mm × 250 mm, 5 μm, 3 mL/min, YMC. Co., Ltd., Tokyo, Japan). Medium-pressure preparation liquid chromatography (MPLC) was performed on a Bona-Agela CHEETAHTM HP100 (Beijing Agela Technologies Co., Ltd., Beijing, China). Column chromatography (CC) was performed with silica gel (200–300 mesh, Qingdao Marine Chemical Inc., Qingdao, China) and Sephadex LH-20 (Amersham Biosciences, San Francisco, CA, USA) [25].

3.2. Fungal Material

The fungal strain FL30r has been previously described [7–10]. The strain was deposited at the Key Laboratory of Marine Drugs, the Ministry of Education of China, School of Medicine and Pharmacy, Ocean University of China, Qingdao, China.

3.3. Fermentation and Extraction

Erlenmeyer flasks (500 mL) containing 150 mL fermentation medium were directly inoculated with spores. The media contained mannitol (20.0 g), glucose (20.0 g), peptone (10.0 g), yeast extract (5.0 g), corn syrup (1.0 g), KH_2PO_4 (0.5 g), and $MgSO_4 \cdot 7H_2O$ (0.3 g) dissolved in 1 L of naturally collected seawater (Huiquan Bay, Yellow sea). The flasks were cultured at 28 °C on a rotary platform shaker at 180 rpm for 9 days. The fermentation broth (45.0 L) was filtered through cheese cloth to separate the supernatant from mycelia. The supernatant was extracted with EtOAc (3 × 45.0 L) and evaporated under reduced pressure to produce the crude gum (15.0 g) [26].

3.4. Isolation

The extract was subjected to VLC (vacuum liquid chromatography) and a stepped gradient elution with petroleum ether/EtOAc (10:0 to 0:10), and EtOAc/MeOH (10:0 to 0:10) was applied to give eight fractions (Fr.1 to Fr.8). Fr.5 was further separated on a Sephadex LH-20 column eluted with MeOH and provided four subfractions (Fr.5-1 to Fr.5-4). Fr.5-3 was then separated by semi-preparative HPLC with MeOH/H_2O (40:60) as a mobile phase to give compound **1** (4.5 mg, t_R 17 min). Fr.6 was further separated by MPLC (C-18 ODS) using a stepped gradient elution with MeOH/H_2O (20:80 to 100:0) to yield five subfractions (Fr.6-1 to Fr.6-5). Fr.6-1 was separated on a Sephadex LH-20 column eluted with MeOH to provide four subfractions (Fr.6-1-1 to Fr.6-1-4). Fr.6-1-3 was further separated by MPLC (C-18 ODS) using a stepped gradient elution with MeOH/H_2O (30:70 to 100:0) to furnish nine subfractions (Fr.6-1-3-1 to Fr.6-1-3-9). Fr.6-1-3-5 was further separated by semi-preparative HPLC eluted with MeOH/H_2O (30:70) to provide compound **3** (7.0 mg t_R 16 min). Fr.6-1-3-9 was further purified by semi-preparative HPLC eluted with MeCN/H_2O (25:75) to give compound **2** (2.7 mg, t_R 16 min). Fr.6-1-4 was further separated by semi-preparative HPLC eluted with MeOH/H_2O (60:40), thus providing compounds **4** (4.4 mg, t_R 13 min) and **5** (2.0 mg, t_R 35 min).

Sorbicillasin A (**1**): pale yellow oil, $[\alpha]_D^{25}$ −15.5 (c 0.15, MeOH); UV (MeOH) λ_{max} (log ε): 220 (2.65), 261 (2.01), 307 (1.80) nm; IR (KBr) ν_{max} 3292, 2906, 1670, 1457, 1362, 1207, 1146, 899 cm^{-1}, see Figure S10; ECD (1 mM MeOH) λ_{max} ($\Delta\varepsilon$) 220 (−3.04), 245 (−1.76), 270 (+2.32) nm; 1D NMR data, see Table 1; HRESIMS *m/z* 375.1550 [M + H]$^+$ (calcd for $C_{19}H_{23}N_2O_6$, 375.1551).

Sorbicillasin B (**2**): pale yellow oil, $[\alpha]_D^{25}$ +10.4 (c 0.15, MeOH); UV (MeOH) λ_{max} (log ε): 220 (2.65), 261 (2.01), 307 (1.80) nm; IR (KBr) ν_{max} 3329, 2926, 1683, 1558, 1384, 1208, 1143, 838 cm^{-1}, see Figure S18; ECD (1 mM MeOH) λ_{max} (Δ_ε) 225 (+1.88), 250 (+4.32), 275 (−1.90) nm; 1D NMR data, see Table 1; HRESIMS *m/z* 397.1376 [M + Na]$^+$ (calcd for $C_{19}H_{22}N_2O_6Na$, 397.1370).

Scirpyrone K (**3**): white amorphous powder; UV (MeOH) λ_{max} (log ε): 231 (1.36), 285 (2.35) nm; IR (KBr) ν_{max} 3104, 2954, 2705, 1743, 1407, 1170, 1001, 836 cm^{-1}, see Figure S26; 1D NMR data, see Table 2; HRESIMS *m/z* 213.0764 [M + H]$^+$ (calcd for $C_{10}H_{13}O_5$, 213.0757).

3.5. Biological Assay

Cytotoxic activities of **1–3** were evaluated using an MGC-803 cell line by the SRB (Sulforhodamine B) method [14] and the K562 cell line by the MTT method [15]. The positive control was doxorubicin hydrochloride. In the DPPH scavenging assay [16], samples to be tested were dissolved in MeOH and the solution (160 µL) was dispensed into wells of a 96-well microtiter tray. Forty microliters of the DPPH solution in MeOH were added to each well. The mixture was shaken and left to stand for 30 min. After the reaction, absorbance was measured at 510 nm, and the percent inhibition was calculated. IC$_{50}$ values denoted the concentration of sample required to scavenge 50% of the DPPH free radicals [27].

3.6. Computation Section

Conformational searches were run by employing the "systematic" procedure implemented in *Spartan'14* [28] using MMFF (Merck molecular force field), which were all reoptimized with DFT calculations at the B3LYP/6-31+G(d) level using the Gaussian09 program [29]. The geometry of various initial conformations was optimized. Vibrational frequency calculations confirmed the presence of minima. Time-dependent DFT calculations were performed on the three lowest-energy conformations for (7*R*,11*S*)-**1a** and three lowest-energy conformations for (7*S*,11*S*)-**1b** (>4% population) using 30 excited states and a polarizable continuum model (PCM) for MeOH. ECD spectra were generated using the program *SpecDis* [30] by applying a Gaussian band shape with 0.30 eV width for **1a** and 0.35 eV width for **1b**, from dipole length rotational strengths. The dipole velocity forms yielded negligible differences. The spectra of the conformers were combined using Boltzmann weighting, with the lowest-energy conformations accounting for 100% of the weights. The calculated spectra were shifted for **1a** (4 nm) and for **1b** (0 nm) to facilitate comparison to the experimental data.

4. Conclusions

In summary, two new nitrogen-containing sorbicillinoids, one new 3,4,6-trisubstituted α-pyrone derivative, and two known biosynthetically related polyketides were isolated from the deep-sea-derived fungus *Phialocephala* sp. FL30r. The absolute configurations of new compounds **1–2** were determined by NMR and TD-DFT calculations of specific ECD spectra. Compounds **1** and **2** were unusual naturally occurring nitrogen-containing sorbicillinoid derivatives with a novel hexahydropyrimido[2,1-*a*] isoindole moiety. Pyrone **3** exhibited radical scavenging activity against DPPH.

Supplementary Materials: The following are available online at http://www.mdpi.com/1660-3397/16/7/245/s1: the HPLC analysis of the metabolic extracts, MS, IR, 1D, and 2D NMR spectra for compounds **1–3**, and details for ECD calculations. Figure S1: HPLC analysis of the metabolic extracts from different conditions; Figures S2–S10: 1D, 2D NMR, NOESY, HRESIMS, and IR spectra of sorbicillasin A (**1**); Figures S11–S18: 1D, 2D NMR, NOESY, HRESIMS, and IR spectra of sorbicillasin B (**2**); Figures S19–S26: 1D, 2D NMR, HRESIMS, and IR spectra of scirpyrone K (**3**); Figure S27: ECD calculation of **1a**; Figure S28: ECD calculation of **1b**.

Author Contributions: The contributions of the respective authors are listed as follows: Z.Z. drafted the work. Z.Z. and X.H. performed the fermentation, extraction, isolation, and structural elucidation of the constituents. Q.C., G.Z., T.Z. and Q.G. contributed to checking and confirming all of the procedures of the isolation and the structural elucidation. D.L. designed the study, supervised the laboratory work, and contributed to the structural determination, bioactivity evaluation and critical reading and revision of the manuscript. All the authors have read the final manuscript and approved the submission.

Funding: This work was financially supported by the NSFC-Shandong Joint Fund for Marine Science Research Centers (U1606403), the Qingdao National Laboratory for Marine Science and Technology (2015ASKJ02 and 2016ASKJ08-02), Shandong province key research and development program (2016GSF201204) and Fundamental Research Funds for the Central Universities (201562016), Qingdao National Laboratory for Marine Science and Technology (No. 2016ASKJ08-02).

Conflicts of Interest: The authors declare no conflict of interest.

References

1. Hoffmeister, D.; Keller, N.P. Natural products of filamentous fungi: Enzymes, genes, and their regulation. *Nat. Prod. Rep.* **2007**, *24*, 393–416. [CrossRef] [PubMed]

2. Sanchez, J.F.; Somoza, A.D.; Keller, N.P.; Wang, C.C.C. Advances in *Aspergillus* secondary metabolite research in the post-genomic era. *Nat. Prod. Rep.* **2012**, *29*, 351–371. [CrossRef] [PubMed]

3. Wang, H.; Eze, P.M.; Hoefert, S.P.; Janiak, C.; Hartmann, R.; Okoye, F.B.C.; Esimone, C.O.; Orfali, R.S.; Dai, H.; Liu, Z.; et al. Substituted L-tryptophan-L-phenyllactic acid conjugates produced by an endophytic fungus *Aspergillus aculeatus* using an OSMAC approach. *RSC Adv.* **2018**, *8*, 7863–7872. [CrossRef]

4. Meng, L.H.; Li, X.M.; Liu, Y.; Xu, G.M.; Wang, B.G. Antimicrobial alkaloids produced by the mangrove endophyte *Penicillium brocae* MA-231 using the OSMAC approach. *RSC Adv.* **2017**, *7*, 55026–55033. [CrossRef]

5. Yuan, C.; Guo, Y.H.; Wang, H.Y.; Ma, X.J.; Jiang, T.; Zhao, J.L.; Zou, Z.M.; Ding, G. Allelopathic polyketides from an endolichenic fungus *Myxotrichum* sp. by using OSMAC strategy. *Sci. Rep. UK* **2016**, *6*, 19350. [CrossRef] [PubMed]

6. Guo, W.; Zhang, Z.; Zhu, T.; Gu, Q.; Li, D. Penicyclones A-E, antibacterial polyketides from the deep sea-derived fungus *Penicillium* sp. F23-2. *J. Nat. Prod.* **2015**, *78*, 2699–2703. [CrossRef] [PubMed]

7. Li, D.; Wang, F.; Xiao, X.; Fang, Y.; Zhu, T.; Gu, Q.; Zhu, W. Trisorbicillinone A, a novel sorbicillin trimer, from a deep ocean sediment derived fungus, *Phialocephala* sp. FL30r. *Tetrahedron Lett.* **2007**, *48*, 5235–5238. [CrossRef]

8. Li, D.; Wang, F.; Cai, S.; Zeng, X.; Xiao, X.; Gu, Q.; Zhu, W. Two new bisorbicillinoids isolated from a deep-sea fungus, *Phialocephala* sp. FL30r. *J. Antibiot.* **2007**, *60*, 317–320. [CrossRef] [PubMed]

9. Li, D.; Cai, S.; Zhu, T.; Wang, F.; Xiao, X.; Gu, Q. Three new sorbicillin trimers, trisorbicillinones B, C, and D, from a deep ocean sediment derived fungus, *Phialocephala* sp. FL30r. *Tetrahedron* **2010**, *66*, 5101–5106. [CrossRef]

10. Li, D.H.; Cai, S.X.; Zhu, T.J.; Wang, F.P.; Xiao, X.; Gu, Q.Q. New cytotoxic metabolites from a deep-sea-derived fungus, *Phialocephala* sp., strain FL30r. *Chem. Biodivers.* **2011**, *8*, 895–901. [CrossRef] [PubMed]

11. Washida, K.; Abe, N.; Sugiyama, Y.; Hirota, A. Novel DPPH radical scavengers, demethylbisorbibutenolide and trichopyrone, from a fungus. *Biosci. Biotechnol. Biochem.* **2007**, *71*, 1052–1057. [CrossRef] [PubMed]

12. Yang, Y.; Yang, F.; Miao, C.; Liu, K.; Li, Q.; Qin, S.; Zhao, L.; Ding, Z. Antifungal metabolites from the rhizospheric *Penicillium* sp. YIM PH 30003 associated with *Panax notoginseng*. *Phytochem. Lett.* **2015**, *11*, 249–253. [CrossRef]

13. Tian, J.F.; Yu, R.J.; Li, X.X.; Gao, H.; Guo, L.D.; Tang, J.S.; Yao, X.S. ^{1}H and ^{13}C NMR spectral assignments of 2-pyrone derivatives from an endophytic fungus of *Sarcosomataceae*. *Magn. Reson. Chem.* **2015**, *53*, 866–871. [CrossRef] [PubMed]

14. Skehan, P.; Storeng, R.; Scudiero, D.; Monks, A.; McMahon, J.; Vistica, D.; Warren, J.T.; Bokesch, H.; Kenney, S.; Boyd, M.R. New colorimetric cytotoxicity assay for anticancer-drug screening. *J. Natl. Cancer Inst.* **1990**, *82*, 1107–1112. [CrossRef] [PubMed]

15. Mosmann, T. Rapid colorimetric assay for cellular growth and survival: Application to proliferation and cytotoxicity assays. *J. Immunol. Methods* **1983**, *65*, 55–63. [CrossRef]

16. Chen, Y.; Wang, M.; Rosen, R.T.; Ho, C.T. 2,2-Diphenyl-1-picrylhydrazyl radical-scavenging active components from *Polygonum multiflorum* Thunb. *J. Agric. Food Chem.* **1999**, *47*, 2226–2228. [CrossRef] [PubMed]

17. Meng, J.; Wang, X.; Xu, D.; Fu, X.; Zhang, X.; Lai, D.; Zhou, L.; Zhang, G. Sorbicillinoids from fungi and their bioactivities. *Molecules* **2016**, *21*, 715. [CrossRef] [PubMed]

18. Bringmann, G.; Lang, G.; Gulder, T.A.M.; Tsuruta, H.; Muhlbacher, J.; Maksimenka, K.; Steffens, S.; Schaumann, K.; Stohr, R.; Wiese, J.; et al. The first sorbicillinoid alkaloids, the antileukemic sorbicillactones A and B, from a sponge-derived *Penicillium chrysogenum* strain. *Tetrahedron* **2005**, *61*, 7252–7265. [CrossRef]

19. Cabrera, G.M.; Butler, M.; Rodriguez, M.A.; Godeas, A.; Haddad, R.; Eberlin, M.N. A sorbicillinoid urea from an intertidal *Paecilomyces marquandii*. *J. Nat. Prod.* **2006**, *69*, 1806–1808. [CrossRef] [PubMed]

20. Guo, W.; Peng, J.; Zhu, T.; Gu, Q.; Keyzers, R.A.; Li, D. Sorbicillamines A-E, nitrogen-containing sorbicillinoids from the deep-sea-derived fungus *Penicillium* sp. F23-2. *J. Nat. Prod.* **2013**, *76*, 2106–2112. [CrossRef] [PubMed]

21. Trifonov, L.S.; Dreiding, A.S.; Hoesch, L.; Rast, D.M. Isolation of four hexaketides from *Verticillium intertextum*. *Helv. Chim. Acta* **1981**, *64*, 1843–1846. [CrossRef]

22. Miklos, F.; Bozo, K.; Galla, Z.; Haukka, M.; Fulop, F. Traceless chirality transfer from a norbornene β-amino acid to pyrimido[2,1-*a*]isoindole enantiomers. *Tetrahedron Asymmetr.* **2017**, *28*, 1401–1406. [CrossRef]

23. Houlihan, W.J.; Kelly, L.; Pankuch, J.; Koletar, J.; Brand, L.; Janowsky, A.; Kopajtic, T.A. Mazindol analogues as potential inhibitors of the cocaine binding site at the dopamine transporter. *J. Med. Chem.* **2002**, *45*, 4097–4109. [CrossRef] [PubMed]

24. Miklos, F.; Toth, Z.; Haenninen, M.M.; Sillanpaeae, R.; Forro, E.; Fueloep, F. Retro-Diels-Alder protocol for the synthesis of pyrrolo[1,2-*a*]pyrimidine and pyrimido[2,1-*a*] isoindole enantiomers. *Eur. J. Org. Chem.* **2013**, *2013*, 4887–4894. [CrossRef]

25. Yu, G.; Wang, S.; Wang, L.; Che, Q.; Zhu, T.; Zhang, G.; Gu, Q.; Guo, P.; Li, D. Lipid-lowering polyketides from the fungus *Penicillium Steckii* HDN13-279. *Mar. Drugs* **2018**, *16*, 25. [CrossRef] [PubMed]

26. Zhang, Z.; He, X.; Zhang, G.; Che, Q.; Zhu, T.; Gu, Q.; Li, D. Inducing secondary metabolite production by combined culture of *Talaromyces aculeatus* and *Penicillium variabile*. *J. Nat. Prod.* **2017**, *80*, 3167–3171. [CrossRef] [PubMed]

27. Guo, W.; Kong, X.; Zhu, T.; Gu, Q.; Li, D. Penipyrols A-B and peniamidones A-D from the mangrove derived *Penicillium solitum* GWQ-143. *Arch. Pharm. Res.* **2015**, *38*, 1449–1454. [CrossRef] [PubMed]

28. *Spartan' 14*; Wavefunction Inc.: Irvine, CA, USA, 2013.

29. Frisch, M.J.; Trucks, G.W.; Schlegel, H.B.; Scuseria, G.E.; Robb, M.A.; Cheeseman, J.R.; Scalmani, G.; Barone, V.; Mennucci, B.; Petersson, G.A.; et al. *Gaussian 09*; Revision A.1; Gaussian, Inc.: Wallingford, CT, USA, 2009.
30. Bruhn, T.; Hemberger, Y.; Schaumlöffel, A.; Bringmann, G. *SpecDis, Version 1.53*; University of Wuerzburg: Wurzburg, Germany, 2011.

Article

Perylenequione Derivatives with Anticancer Activities Isolated from the Marine Sponge-Derived Fungus, *Alternaria* sp. SCSIO41014

Xiaoyan Pang [1,2], Xiuping Lin [1], Pei Wang [3], Xuefeng Zhou [1], Bin Yang [1], Junfeng Wang [1,*] and Yonghong Liu [1,2,*]

[1] CAS Key Laboratory of Tropical Marine Bio-resources and Ecology, Guangdong Key Laboratory of Marine Materia Medica, RNAM Center for Marine Microbiology, South China Sea Institute of Oceanology, Chinese Academy of Sciences, Guangzhou 510301, China; luckygirlpxy@163.com (X.P.); xiupinglin@hotmail.com (X.L.); xfzhou@scsio.ac.cn (X.Z.); yangbin@scsio.ac.cn (B.Y.)

[2] University of Chinese Academy of Sciences, Beijing 100049, China

[3] Key Laboratory of Biology and Genetic Resources of Tropical Crops, Ministry of Agriculture, Institute of Tropical Bioscience and Biotechnology, Chinese Academy of Tropical Agricultural Sciences, Haikou 571101, China; wangpei@itbb.org.cn

[*] Correspondence: wangjunfeng@scsio.ac.cn (J.W.); yonghongliu@scsio.ac.cn (Y.L.); Tel.: +86-020-8902-3174 (J.W.); +86-020-8902-3244 (Y.L.)

Received: 25 July 2018; Accepted: 8 August 2018; Published: 14 August 2018

Abstract: Seven new secondary metabolites classified as two perylenequinone derivatives (**1** and **2**), an altenusin derivative (**3**), two phthalide racemates (**4** and **5**), and two phenol derivatives (**6** and **7**), along with twenty-one known compounds (**8–28**) were isolated from cultures of the sponge-derived fungus, *Alternaria* sp. SCSIO41014. The structures and absolute configurations of these new compounds (**1–7**) were determined by spectroscopic analysis, X-ray single crystal diffraction, chiral-phase HPLC separation, and comparison of ECD spectra to calculations. Altertoxin VII (**1**) is the first example possessing a novel 4,8-dihydroxy-substituted perylenequinone derivative, while the phenolic hydroxy groups have commonly always substituted at C-4 and C-9. Compound **1** exhibited cytotoxic activities against human erythroleukemia (K562), human gastric carcinoma cells (SGC-7901), and hepatocellular carcinoma cells (BEL-7402) with IC_{50} values of 26.58 ± 0.80, 8.75 ± 0.13, and 13.11 ± 0.95 µg/mL, respectively. Compound **11** showed selectively cytotoxic activity against K562, with an IC_{50} value of 19.67 ± 0.19 µg/mL. Compound **25** displayed moderate inhibitory activity against *Staphylococcus aureus* with an MIC value of 31.25 µg/mL.

Keywords: sponge-derived fungus; *Alternaria* sp.; perylenequinone derivatives; X-ray single crystal diffraction; cytotoxic activity; antibacterial

---

1. Introduction

Perylenequinone derivatives are secondary metabolites characterized by a conjugated aromatic pentacyclic dione, which are mainly derived from fungi [1,2]. Hypocrellins, which are the typical perylenequinone derivatives isolated from fungi *Hypocrella bambusae* and *Shiraia bambusicola*, have been studied for their light-induced antitumor and antiviral activities [3]. Furthermore, many perylenequinone derivatives isolated from *Alternaria* sp. showed phytotoxicity, as well as antimicrobial and anticancer activities [4–7]. Sponge-derived fungi are one of the richest sources of many structurally unique and biologically active secondary metabolites among marine sources [8]. As part of our ongoing research for bioactive natural products from sponge-derived fungi [9–13], the fungus *Alternaria* sp. SCSIO41014 was studied. Seven new (**1–7**) and twenty-one known (**8–28**) compounds (Figure 1) were isolated from the culture extract of the fungus *Alternaria* sp. SCSIO41014.

Altertoxin VII (**1**) is the first example possessing a novel 4,8-dihydroxy-substituted perylenequinone derivative, while the phenolic hydroxy groups have always commonly substituted at C-4 and C-9. Herein, we describe the structure elucidation and bioactivity evaluation of new compounds **1**–**7**.

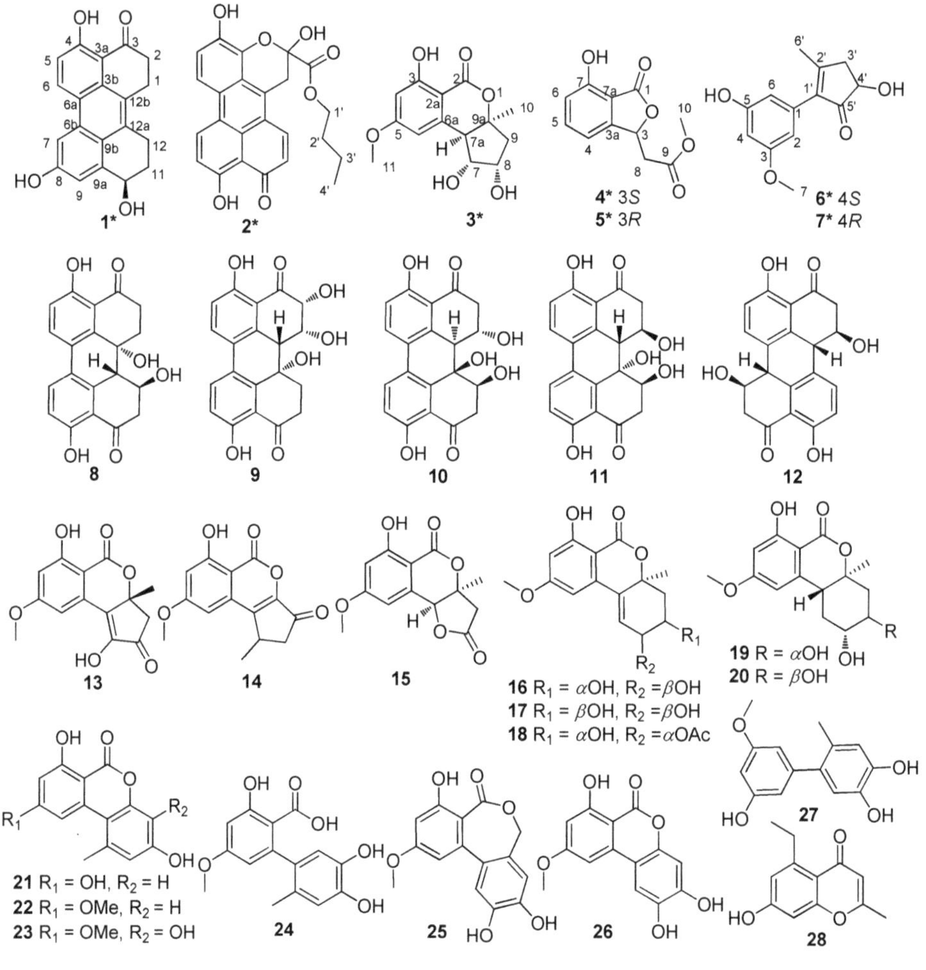

Figure 1. Chemical structures of compounds **1**–**28**.

2. Results and Discussion

2.1. Structural Elucidation

Compound **1** was obtained as a dark red powder. Its molecular formula was established as $C_{20}H_{16}O_4$ by ^{13}C NMR data and the high resolution electrospray ionization mass spectroscopy (HRESIMS) $[M + Na]^+$ peak at *m*/*z* 343.0939 (calculated for $C_{20}H_{15}O_4Na$, 343.0941), indicating thirteen degrees of unsaturation. Its UV spectrum showed maxima at 323 and 368 nm, which suggested that **1** featured a polycyclic aromatic hydrocarbons system. Its ^1H NMR data (Table 1) showed two *ortho* aromatic protons (δ_H 8.79, d, *J* = 9.5 Hz, H-6 and 7.16, d, *J* = 9.0 Hz, H-5), two *meta* aromatic protons (δ_H 7.82, d, *J* = 2.0 Hz, H-7 and 7.30, d, *J* = 1.5 Hz, H-9), an oxygenated methine (δ_H 4.82, dd, *J* = 9.0, 3.0 Hz, H-10), and three hydroxy protons (δ_H 13.27, brs, OH-4; 9.87, brs, OH-8; 5.48, brs,

OH-10). The striking downfield shift of OH-4 is a typical feature of a strong hydrogen bond, which is in accordance with the downfield shift of C-3 (δ_C 204.0) observed in the ^{13}C NMR spectrum. The ^{13}C NMR (DEPT) (Table 1) data also exhibited twenty carbon signals, including four sp^3 methylenes, one oxygenated sp^3 methine (δ_C 67.4, C-10), four sp^2 methines (δ_C 133.1, C-6; 116.2, C-5; 114.4, C-9; 104.4, C-7), ten sp^2 non-protonated carbons, and one carbonyl. Besides the presence of two benzene rings, a double bond, and a carbonyl, three degrees of unsaturation were left, which indicated that there were another three rings in the structure of **1**. Half of its NMR data were identical to those of 4,9-dihydroxy-1,2,11,12-tetrahydroperylene-3,10-quinone [14], while compound **1** was not a symmetric structure, and the other half of the data had two obvious differences. One carbonyl and two *ortho* aromatic protons in 4,9-dihydroxy-1,2,11,12-tetrahydroperylene-3,10-quinone were replaced by an oxygenated methine and two *meta* aromatic protons, respectively. The differences were testified by the cross-peaks of H$_2$-1 (δ_H 3.25)/H$_2$-2 (δ_H 2.92), of H-5/H-6, and of OH-10/H-10/H$_2$-11 (δ_H 2.18 and 1.88)/H$_2$-12 (δ_H 3.20 and 3.00) in the ^{1}H-^{1}H correlation spectroscopy (COSY) spectrum (Figure 2), and the correlations from: H-7 and H-9 to C-8; from H-10 and H$_2$-11 to C-9a; from H-7, H-9, H-10 and H$_2$-12 to C-9b; from H-9 to C-7 and C-10; and from H$_2$-11 and H$_2$-12 to C-12a in the heteronuclear multiple bond correlation (HMBC) spectrum (Figure 2). Furthermore, it is allowed for the linkage of ring A (Figure 2) by the correlations from H$_2$-1 to C-12a; from H-6 to C-6b; from H-7 to C-6a; and from H$_2$-12 to C-12b in the HMBC spectrum. The attachment of the three hydroxy groups was further confirmed by the correlations from: OH-4 to C-3a, C-4 and C-5; from OH-8 to C-7, C-8 and C-9; and from OH-10 to C-9a (Figure 2). Therefore, the planar structure of **1** was established as 4,8,10-trihydroxy-1,2,11,12-tetrahydroperylene-3-quinone. Its nuclear Overhauser effect spectroscopy (NOESY) spectrum displayed cross-peaks of H$_2$-1/H$_2$-12 and of H-6/H-7, which also support the conjectural structure above. The absolute configuration of **1** was established based on comparison of its experimental electronic circular dichroism (ECD) curve with the calculated ECD curve of the 10*R* and the 10*S* model at the B3LYP/6-31G(d,p) level in Gaussian 03, and the former was in accordance with the experimental one (Figure 3). Thus, the absolute structure of **1** was defined as (10*R*)-4,8,10-trihydroxy-1,2,11,12-tetrahydroperylene-3-quinone, and named altertoxin VII.

Table 1. ^{1}H NMR and ^{13}C NMR data for compounds **1** and **2** in DMSO-d_6 (500, 125 MHz).

No.	δ_C, type	δ_H (J in Hz)	No.	δ_C, type	δ_H (J in Hz)
	1			**2**	
1	23.2, CH$_2$	3.25 td (7.0, 2.0)	1	33.9, CH$_2$	4.03 d (16.5) 3.91 d (17.0)
2	36.4, CH$_2$	2.92 t (7.5)	2	95.5, C	
3	204.8, C		3		
3a	110.7, C		3a	137.8, C	
3b	131.6, C		3b	118.3, C	
4	161.9, C		4	142.8, C	
5	116.2, CH	7.16 d (9.0)	5	122.2, CH	7.51 d (9.0)
6	133.1, CH	8.79 d (9.5)	6	115.6, CH	8.35 d (9.0)
6a	120.7, C		6a	124.4, C	
6b	130.8, C		6b	120.3, C	
7	104.4, CH	7.82 d (2.0),	7	133.5, CH	9.08 d (9.5)
8	156.0, C		8	118.0, CH	7.40 d 9.0
9	114.4, CH	7.30 d (1.5)	9	165.6, C	
9a	142.1, C		9a	111.3, C	
9b	120.1, C		9b	124.6, C	
10	67.4, CH	4.82 dd (9.0, 3.0)	10	188.2, C	
11	31.5, CH$_2$	2.18 dq (10.0, 4.5) 1.88 dtd (12.5, 9.5, 4.5)	11	125.9, CH	6.94 d (10.0)
12	24.3, CH$_2$	3.20 dt (17.0, 5.5) 3.00 ddd (16.5, 10.0, 5.0)	12	139.0, CH	8.63 d (10.0)
12a	133.4, C		12a	119.9, C	
12b	121.4, C		12b	136.5, C	
OH-4		13.27 brs	13	168.6, C	
OH-8		9.87 brs	1'	65.1, CH$_2$	4.09 td (6.5, 2.0)
OH-10		5.48 brs	2'	30.0, CH$_2$	1.46 qui (7.5)
			3'	18.3, CH$_2$	1.14 sex (7.5)
			4'	13.4, CH$_3$	0.71 t (7.5)
			OH-2		9.76 brs
			OH-4		8.14 brs
			OH-9		15.12 brs

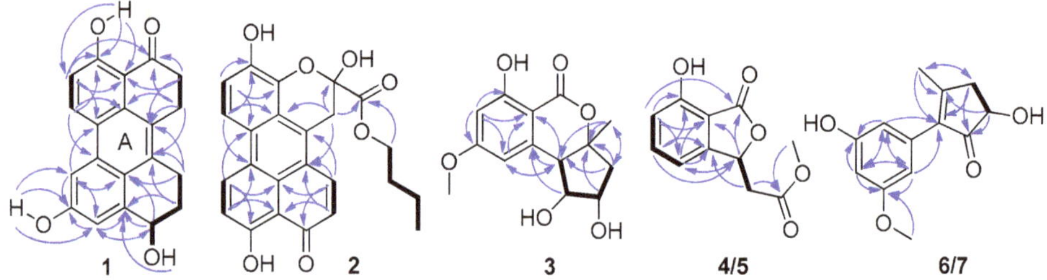

Figure 2. COSY "H—H" and key HMBC "H→C" correlations of compounds **1–7**.

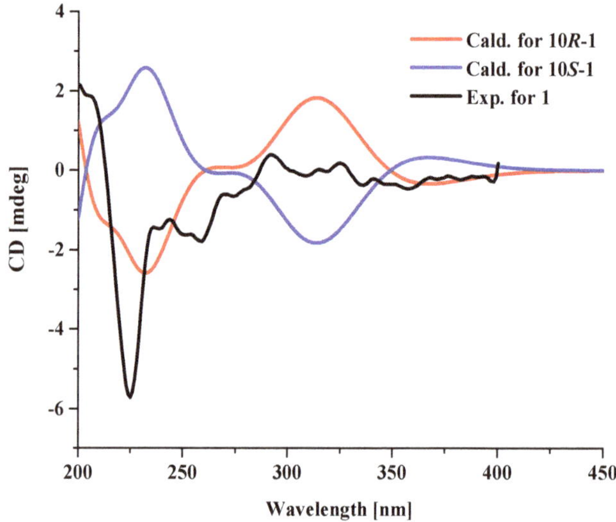

Figure 3. Comparison between calculated and experimental electronic circular dichroism (ECD) spectra of compound **1**.

Compound **2** was obtained as a dark red powder. Its molecular formula was assigned as $C_{24}H_{20}O_7$ based on its $[M + Na]^+$ ion at m/z 443.1110 and $[M + H]^+$ ion at m/z 421.1284 in the HRESIMS spectrum, indicating fifteen degrees of unsaturation. Analysis of its ^1H NMR and ^{13}C NMR data revealed that the structural features of **2** were similar to those of xanalteric acid II [5], except for the presence of an oxygenated *n*-butyl group ($\delta_{C/H}$ 65.1/4.09, CH_2-1′; 30.0/1.46, CH_2-2′; 18.3/1.14, CH_2-3′; 13.4/0.71, CH_3-4′), which was confirmed by its COSY cross-peaks of H_2-1′/H_2-2′/H_2-3′/H_3-4′. The location of the oxygenated *n*-butyl group was ascertained by the HMBC correlation from H_2-1′ to C-13. Based on the HMBC correlations (Figure 2), the planar structure of **2** was determined as butyl 2,4,9-trihydroxy-10-oxo-2,10-dihydro-1*H*-phenaleno[1,2,3-*de*] chromene-2-carboxylate and named butyl xanalterate. Due to the existence of the cyclic hemi-ketal, compound **2** was not stable under the protic solvent condition, meaning that the absolute configuration was uncertain.

Compound **3** possessed a molecular formula of $C_{14}H_{16}O_6$ on the basis of its NMR data and the HRESIMS ion peak at m/z 303.0844 $[M + Na]^+$, indicating seven indices of hydrogen deficiency. The ^1H NMR spectrum (Table 2) exhibited two aromatic protons (δ_H 6.52, brs, H-6; 6.43, brs, H-4); three methines (δ_H 4.20, ddd, J = 7.0, 5.5, 2.5 Hz, H-8; 3.90, dd, J = 10.5, 5.0 Hz, H-7; 3.07, d, J = 10.0 Hz, H-7a); one methoxy group (δ_H 3.88, s, H_3-11); and a methyl singlet (δ_H 1.47, s, H_3-10). The ^{13}C NMR and HSQC spectra indicated the presence of 14 carbons, including two methyls (δ_C 56.2, C-11; 25.7, C-10), one sp^3 methylene, three sp^3 methines (two oxygenated at δ_C 80.1, C-7; 71.2, C-8), two sp^2 methines (δ_C 101.0, C-4; 108.9, C-6), four sp^2 non-protonated carbon, and an ester carbonyl (δ_C 170.0, C-2). Because one benzene ring and one carbonyl accounted for five degrees of unsaturation, double ring fragment was required for the structure of **3**. Its NMR data were similar to those of dihydroaltenuenes A (**19**) [15], the obvious differences being the disappearance of a methylene signal (δ_C 28.1) and the downfield shift of the oxygenated methine in **3**, which indicated that a fragment of the six-membered ring in **19** was replaced by a five-membered ring in **3**. The deduction was confirmed by following COSY correlations of H-7a/H-7/H-8/H_2-9, as well as HMBC correlations, from H-7 and H-7a to C-6a; from H-6, H_2-9 and H_3-10 to C-7a; and from H-8, H_2-9 and H_3-10 to C-9a (Figure 2). The NOESY spectrum in DMSO-d_6 (Figure 4) exhibited the cross-peaks of H-7a to H-9α, H_3-10, OH-7, and OH-8, and also of H-9β to H-7, which suggested that CH_3-10, H-7a, OH-7, and OH-8 were on the same side. The Cu Kα radiation for the X-ray diffraction experiment with the refined Flack parameter of $-0.02(7)$ allowed the assignment of the absolute configuration of all the stereogenic centers in **3** as 7*R*, 7a*R*, 8*S*, and 9a*S*. Thus, the absolute structure of **3** was unambiguously elucidated and named nordihydroaltenuenes A.

Table 2. ^{1}H NMR and ^{13}C NMR data for compounds **3–7** in CD$_3$OD (500, 125 MHz).

No.	3 δ_C, type	δ_H (*J* in Hz)	No.	4/5 δ_C, type	δ_H (*J* in Hz)	No.	6/7 δ_C, type	δ_H (*J* in Hz)
2	170.0, C		1	171.2, C		1	134.6, C	
2a	100.1, C		3	78.3, CH	5.80 (dd, 8.0, 4.5)	2	109.8, CH	6.31 overlap
3	167.8, C		3a	152.1, C		3	162.2, C	
4	101.0, CH	6.43 brs	4	113.9, CH	6.99 (d, 7.5)	4	101.9, CH	6.35 t (2.5)
5	167.8, C		5	137.8, CH	7.52 (t, 8.0)	5	159.5, C	
6	108.9, CH	6.52 brs	6	117.0, CH	6.88 (d, 8.0)	6	107.2, CH	6.32 overlap
6a	143.1, C		7	158.2, C		7	55.7, CH$_3$	3.75 s
7a	51.8, CH	3.07 d (10.0)	7a	112.4, C		1′	139.1, C	
7	80.1, CH	3.90 dd (10.5, 5.0)	8	40.0, CH$_2$	2.76 dd (16.5, 8.0) 3.07 dd (17.0, 4.5)	2′	171.9, C	
8	71.2, CH	4.20 ddd (7.0, 5.5, 2.5)	9	171.6, C		3′	41.9, CH$_2$	3.05 dd (18.0, 6.5) 2.50 ddd (18.0, 3.0, 1.5)
9β 9α	47.8, CH$_2$	2.60 dd (15.5, 7.0) 2.04 dd (15.5, 1.5)	10	52.5, CH$_3$	3.69 s	4′	72.6, CH	4.30 dd (7.0, 3.0)
9a	89.4, C					5′	208.8, C	
10	25.7, CH$_3$	1.47 s				6′	18.5, CH$_3$	2.18 s
11	56.2, CH$_3$	3.88 s						

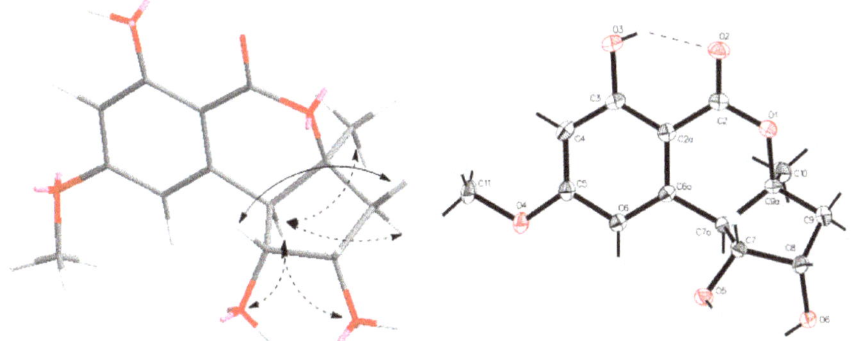

Figure 4. Key NOESY correlations and ORTEP drawing of compound **3**.

Compounds **4** and **5** were initially believed to be a single compound, which was obtained as a yellow oil. The molecular formula was established as $C_{11}H_{15}O_5$ based on the HRESIMS and ^{13}C NMR data. A literature survey suggested that the ^{1}H NMR and ^{13}C NMR data (Table 2) closely resembled those of isoochracinic acid [16], except for the presence of a methoxy group ($\delta_{C/H}$ 52.5/3.69, C-10) and an upfield shift of 1.3 ppm of the carbonyl (δ_C 171.6, C-9), which indicated the carboxyl in the isoochracinic acid was methylated. The deduction was confirmed by the HMBC correlation of H_3-10 to C-9. The planar structure of **4/5** was further established by the COSY and HMBC correlations (Figure 2). The optical rotation was small ($[\alpha]_D^{25}$ +1.6) and the value of the ECD spectrum close to zero, suggesting that compounds **4** and **5** belonged to a racemate. Using a chiral-phase column (Daicel Chiraloak IC-3, 250 × 4.6 mm, 5 μm), the racemate was resolved to two enantiomers, **4** and **5**, whose scale was nearly 1:1 (Figure S30). The optical rotation of **4** ($[\alpha]_D^{25}$ −21, *c* 0.1, MeOH) and **5** ($[\alpha]_D^{25}$ +26, *c* 0.1, MeOH), as well as the ECD spectrum (Figure 5), were opposite to each other. In general, the 3*R*-phthalides displayed a positive optical rotation, while 3*S*-phthalides had a negative one [17–20]. Therefore, the absolute configuration of compound **4** was proposed as 3*S* and named (*S*)-isoochracinate A1, while compound **5** was proposed as 3*R*, and named (*R*)-isoochracinate A2.

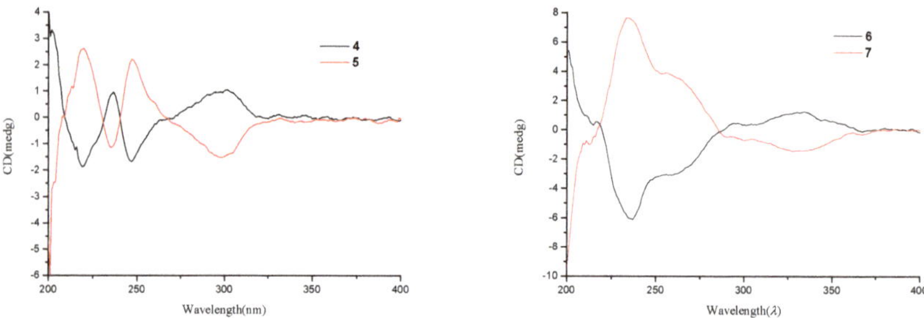

Figure 5. Experimental ECD spectra of compounds **4–7**.

Compounds **6** and **7**, obtained as a yellow powder, were also believed to be a single compound at first. The HRESIMS spectrum supported a molecular formula of $C_{13}H_{14}O_4$, requiring 7 degrees of unsaturation. The ^{1}H NMR and ^{13}C NMR data (Table 2) of **6/7** were almost identical to those of 4′-(*S*)-(3,5-dihydroxyphenyl)-4′-hydroxy-6′-methylcyclopent-1′-en-5′-one isolated from *Penicillium* sp. HN29-3B1 [21], indicating that **6/7** were a structural analog, except for the presence of a methoxy group ($\delta_{C/H}$ 55.7, C-7). The deduction was confirmed by the HMBC correlation (Table 2) from H$_3$-7 to C-3. Analyzed by a chiral-phase column (Phenomenex Lux Cellulose-2, 250 × 4.6 mm, 5 μm), compounds **6/7** were found to be stereoisomeric mixture, whose mass ratio was about 1:3.5 (Figure S39). The single compound **6** ($[\alpha]_D^{25}$ +3.1, *c* 0.1, MeOH) and **7** ($[\alpha]_D^{25}$ −7.2, *c* 0.1, MeOH) were isolated, and they had opposite ECD spectra (Figure 5). The optical rotation of **6** was consistent with that of 4′-(*S*)-(3,5-dihydroxyphenyl)-4′-hydroxy-6′-methylcyclopent-1′-en-5′-one [21], indicating that they had the same *S* configuration at C-4′, opposite to that of **7**. Thus, the absolute structure of **6** was determined as 4′-(*S*)-(3-Methoxy-5-hydroxyphenyl)-4′-hydroxy-6′-methylcyclopent-1′-en-5′-one and named (*S*)-alternariphent A1, while the absolute structure of **7** was determined as 4′-(*R*)-(3-Methoxy-5-hydroxyphenyl)-4′-hydroxy-6′-methylcyclopent-1′-en-5′-one and named (*R*)-alternariphent A2.

By comparing NMR data with the values in literature, the structures of the twenty-one known compounds were identified as altertoxin I (**8**) [22], 7-epi-8-hydroxyaltertoxin (**9**) [6], stemphytriol (**10**) [23], 6-epi-stemphytriol (**11**) [6], stemphyperylenol (**12**) [23,24], (*R*)-1,6-dihydroxy-8-methoxy-3a-methyl-3,3a-dihydrocyclopenta[*c*]iso- chromene-2,5-dione (**13**) [25], 1-deoxyrubralactone (**14**) [25], 6-hydroxy-8-methoxy-3a-methyl-3a,9b-dihydro-3*H*-furo[3,2-c]isochromene -2,5-dione (**15**) [26], altenuene (**16**) [27], 4′-epialtenuene (**17**) [28], (−)-(2*R*,3*R*,4a*R*)-altenuene-3-acetoxy ester (**18**) [29], dihydroaltenuenes A (**19**) [15], 3-epi-dihydroaltenuene A (**20**) [30], alternariol (**21**) [31], alternariol monomethyl ether (**22**) [32], 3′-hydroxyalternariol-5-*O*-methyl ether (**23**) [28], altenusin (**24**) [33], alterlactone (**25**) [28], altenuisol (**26**) [34], 5′-methoxy-6-methyl-biphenyl-3,4,3′-triol (**27**) [26], and 2,5-dimethyl-7-hydroxychromone (**28**) [35], respectively.

2.2. Biological Activity

The cytotoxic activities of compounds **1**, **2**, and **8–12** against human erythroleukemia (K562), human gastric carcinoma cells (SGC-7901), and hepatocellular carcinoma cells (BEL-7402) were evaluated using the CCK-8 method, and paclitaxel was used as a positive control with IC_{50} values of 0.18 ± 0.20, 0.89 ± 0.15, and 0.54 ± 0.20 µg/mL, respectively. Among the tested compounds, compound **1** exhibited cytotoxic activity against the K562, SGC-7901, and BEL-7402 cell lines with IC_{50} values of 26.58 ± 0.80, 8.75 ± 0.13, and 13.11 ± 0.95 µg/mL, respectively. Compound **11** showed selectively cytotoxic activity against K562 with an IC_{50} value of 19.67 ± 0.19 µg/mL. All compounds were tested for their antibacterial activities against *Staphylococcus aureus*. Compounds **10** and **25** with 50 µg/disc displayed an inhibition zone with a diameter of about 21 and 15 mm, respectively (Figure S1). Furthermore, their minimum inhibitory concentrations (MIC) were tested, and the MIC value of compound **25** was 31.25 µg/mL, while compound **10** showed more than 500 µg/mL—perhaps due to its poor solubility. Ampicillin was used as a positive control with an MIC value of 6.25 µg/mL.

3. Materials and Methods

3.1. General Experimental Procedures

HRESIMS data were recorded on a maXis Q-TOF mass spectrometer in a positive ion mode (Bruker, Fällanden, Switzerland). 1D and 2D NMR spectra were measured on an AV 500 MHz NMR spectrometer (Bruker, Fällanden, Switzerland) with TMS as an internal standard. Chemical shifts were given as δ values, with J values reported in Hz. Optical rotations were measured using a MCP-500 polarimeter (Anton, Austria). UV spectra were recorded on a UV-2600 UV-Vis spectrophotometer (Shimadzu, Japan). X-ray diffraction intensity data were collected on an XtalLAB PRO single-crystal diffractometer using Cu Kα radiation (Rigaku, Japan). ECD spectrum was measured with a Chirascan circular dichroism spectrometer (Applied Photophysics, Surrey, UK). HPLC was performed on a Hitachi Primaide with the YMC ODS SERIES column (YMC-Pack ODS-A, YMC Co. Ltd., Kyoto, Japan, 250×10 mm I.D., S-5 µm, 12 nm) and chiral-phase column (Phenomenex Lux Cellulose-2 column, 4.6 mm $\times$ 25 mm and Daicel Chiraloak IC-3 column, 4.6 mm $\times$ 25 mm). Column chromatography (CC) was carried out on silica gel (200–300 mesh, Jiangyou Silica Gel Development Co., Yantai, China), Sephadex LH-20 (40–70 µm, Amersham Pharmacia Biotech AB, Uppsala, Sweden), and YMC Gel ODS-A (12 nm, S-50 µm YMC Co. Ltd., Kyoto, Japan). Spots were detected on TLC under UV light or by heating after spraying with the mixed solvent of saturated vanillin and 5% H_2SO_4 in EtOH. The TLC plates with silica gel GF254 (0.4–0.5 mm, Qingdao Marine Chemical Factory, Qingdao, China) were used for analysis and for the preparatives.

3.2. Fungal Material

The fungal strain SCSIO41014 was obtained from a *Callyspongia* sp. sponge, which was collected from the sea area near Xuwen County, Guangdong Province, China. The producing strain was stored on MB agar (malt extract 15 g, agar 16 g, sea salt 10 g, water 1 L, pH 7.4–7.8) slants at 4 °C and deposited at the Key Laboratory of Tropical Marine Bio-resources and Ecology, Chinese Academy of Science. The ITS1-5.8S-ITS2 sequence region (508 base pairs, GenBank accession No. MH444654) of strain SCSIO41014 was amplified by PCR, and DNA sequencing showed it shared a significant homology to several species of *Setosphaeria*. The 508 base pairs of the ITS sequence had a 99% sequence identity to that of the *Alternaria alternata* strain SCAU091 (GenBank accession No. MF061753.1). It was thus designated as a member of *Alternaria* sp. and named *Alternaria* sp. SCSIO41014.

3.3. Fermentation and Extraction

The strain SCSIO41014 was cultured in 100 mL flasks ($\times 59$) each containing 10 mL seed medium (malt extract: 15 g, sea salt: 2.5 g, distilled water: 1 L, pH 7.4–7.8) at 27 °C on a rotary shaker (172 rpm) for 48 h. The mass fermentation of this fungus was carried out at 25 °C for 32 days using a rice medium (rice: 200 g/flask, sea salt: 2.5 g/flask, tap water: 200 mL/flask) in the 1 L flasks ($\times$59). The flasks were incubated statically at 25 °C under the normal day and night cycle. After 32 days, cultures were soaked in acetone (400 mL/flask), mashed into small pieces, and vibrated with ultrasound for 20 min. Then the acetone solution was evaporated under reduced pressure to afford an aqueous solution, which was extracted with ethyl acetate (EtOAc) three times. Concurrently, the rice residue was extracted with EtOAc in order to make another EtOAc solution. Both of the EtOAc solutions were combined and concentrated under reduced pressure to produce a crude extract. The extract was then suspended in MeOH and partitioned with equal volumes of petroleum ether (PE). Finally, the MeOH solution was concentrated under reduced pressure to obtain a reddish-brown extract (78.0 g).

3.4. Isolation and Purification

The reddish-brown extract was subjected to silica gel CC, which was eluted with a CH_2Cl_2 and MeOH mixed solvent in a step gradient (100:0–3:1, v/v) and separated into seven fractions (Fr-1–Fr-7). Fr-1 (3.4 g) was subjected to reversed-phase C-18 MPLC with MeOH/H_2O (35:65–100:0, v/v) to get four fractions (Fr-1-1–Fr-1-4). Fr-1-1 was directly separated by semi-preparative HPLC (75% MeOH/H_2O, 1.6 mL/min) to produce **22** (5.4 mg, t_R = 25 min). Fr-1-2 was purified by semi-preparative HPLC (42% CH_3CN/H_2O, 2 mL/min) to yield **13** (26.0 mg, t_R = 22.0 min) and **14** (5.6 mg, t_R = 34.0 min). Fr-1-3 was separated by semi-preparative HPLC (30% CH_3CN/H_2O, 2 mL/min) to yield **15** (17.8 mg, t_R = 43.0 min). Fr-1-4 was purified by semi-preparative HPLC (28% CH_3CN/H_2O, 2 mL/min) to afford raceme **4/5** (97.3 mg, t_R = 18.0 min), part of which was further separated by HPLC with a chiral-phase column (Daicel Chiraloak IC-3 column, 4.6 mm $\times$ 25 mm, eluent *n*-hexane/*iso*-propanol, 65:35 v/v, 1 mL/min) to obtain **4** (3.3 mg, t_R = 15.6 min) and **5** (2.8 mg, t_R = 17.0 min). Fr-3 (15.8 g) was subjected to silica gel CC eluted with a PE and acetone mixed solvent in a step gradient (10:1–0:1, v/v) to gain five fractions (Fr-3-1–Fr-3-5). Fr-3-3 (0.94 g) was subjected to a Sephadex LH-20 column, eluted with MeOH and further purified by semi-preparative HPLC, to produce **18** (7.6 mg, 40% CH_3CN/H_2O, 2 mL/min, t_R = 30.0 min), **23** (8.0 mg, 45% CH_3CN/H_2O, 2 mL/min, t_R = 22.0 min), and **28** (22.5 mg, 45% CH_3CN/H_2O, 2 mL/min, t_R = 24.0 min). Fr-3-4 (7.3 g) was subjected to reversed-phase C-18 MPLC with MeOH/H_2O (10:90–100:0, v/v) to get three fractions (Fr-3-4-1–Fr-3-4-3). Fr-3-4-1 was subjected to a Sephadex LH-20 column eluted with MeOH to gain four parts. One part was purified by a preparative thin-layer chromatography, using CH_2Cl_2/MeOH (10:1) as a developing solvent to afford **16** (42.6 mg, Rf = 0.6) and **19** (11.3 mg, Rf = 0.7). Other parts were separated by semi-preparative HPLC to yield **3** (3.5 mg, 60% MeOH/H_2O, 2 mL/min, t_R = 13.5 min), **9** (4.8 mg, 60% MeOH/H_2O, 2 mL/min, t_R = 15.8 min), **10** (10.3 mg, 60% MeOH/H_2O, 2 mL/min, t_R = 19.0 min), and **25** (104.0 mg, 34% CH_3CN/H_2O, 2 mL/min, t_R = 20.0 min). Fr-3-4-2 was subjected to a Sephadex LH-20 column, eluted with MeOH and purified by semi-preparative HPLC, to yield **11** (5.2 mg, 56% MeOH/H_2O, 2 mL/min, t_R = 26.0 min), **12** (13.1 mg, 35% CH_3CN/H_2O, 2 mL/min, t_R = 21.4 min), and **21** (20.6 mg, 40% CH_3CN/H_2O, 2 mL/min, t_R = 21.0 min). Fr-3-4-3 was subjected to silica gel CC with a PE and EtOAc mixed solvent in a step gradient (5:1–1:1, v/v), and then purified by semi-preparative HPLC to yield **26** (7.0 mg, 30% CH_3CN/H_2O, 2 mL/min, t_R = 26.4 min). Fr-3-5 (1.1 g) was subjected to a Sephadex LH-20 column eluted with MeOH, and then purified by semi-preparative HPLC to yield **2** (6.7 mg, 55% CH_3CN/H_2O, 2 mL/min, t_R = 32.0 min) and **8** (3.9 mg, 48% CH_3CN/H_2O, 2 mL/min, t_R = 18.0 min). Fr-4 (8.4 g) was subjected to silica gel CC eluted with a PE and acetone mixed solvent in a step gradient (10:1–0:1, v/v) to gain three fractions (Fr-4-1–Fr-4-3). Fr-4-2 (5.1 g) was subjected to a Sephadex LH-20 column eluted with MeOH, then subjected to reversed-phase C-18 MPLC with MeOH/H_2O (10:90–100:0, v/v), and further purified by semi-preparative HPLC, to obtain **1** (17.4 mg, 38% CH_3CN/H_2O, 2 mL/min, t_R = 28.2 min), **17** (34.7 mg, 56% MeOH/H_2O, 2 mL/min, t_R = 21.5 min), **20** (10.6 mg, 28% CH_3CN/H_2O, 2 mL/min, t_R = 13.0 min), and the stereoisomeric mixture **6/7** (13.2 mg, 15% CH_3CN/H_2O, 2 mL/min, t_R = 23.6 min). Part of the mixture **6/7** was separated by HPLC with a chiral-phase column (Phenomenex Lux Cellulose-2, 4.6 mm $\times$ 25 mm, eluent *n*-hexane/*iso*-propanol, 40:60 v/v, 1 mL/min) to afford **6** (1.5 mg, t_R = 7.4 min) and **7** (5.2 mg, t_R = 8.4 min). Fr-5 (13.1 g) was subjected to silica gel CC eluted with a PE and acetone mixed solvent in a step gradient (10:1–0:1, *v/v*) to gain three fractions (Fr-5-1–Fr-5-3). Fr-5-2 (3.7 g) was subjected to a Sephadex LH-20 column eluted with MeOH, which was then subjected to reversed-phase C-18 MPLC with MeOH/H_2O (10:90–100:0, *v/v*), and further purified by semi-preparative HPLC, to yield **27** (8.2 mg, 52% MeOH/H_2O, 2 mL/min, t_R = 17.6 min) and **24** (21.9 mg, 52% MeOH/H_2O, 2 mL/min, t_R = 28.0 min).

3.5. Spectral Data

Altertoxin VII (**1**): dark red powder; $[\alpha]_D^{25}$ −9.0 (*c* 0.02, MeOH); UV(MeOH) λ_{max}(log ε) 226 (3.06), 253 (3.36), 323 (2.68), 368 (2.45) nm; ECD (0.38 mM, MeOH) λ_{max} (Δε) 204 (+0.92), 224 (−2.77) and 253 (−0.87) nm; ^{1}H NMR (DMSO-d_6, 500 MHz) and ^{13}C NMR (DMSO-d_6, 125 MHz), Table 1; HRESIMS *m/z* 343.0939 [M + Na]$^+$ (calcd for $C_{20}H_{16}NaO_4$, 343.0941).

Butyl xanalterate (**2**): dark red powder; $[\alpha]_D^{25}$ −42 (*c* 0.02, MeOH); ^{1}H NMR (DMSO-d_6, 500 MHz) and ^{13}C NMR (DMSO-d_6, 125 MHz), Table 1; HRESIMS *m/z* 443.1110 [M + Na]$^+$ (calcd for $C_{24}H_{20}NaO_7$, 443.1101) and 421.1284 [M + H]$^+$ (calcd for $C_{24}H_{21}O_7$, 421.1282).

Nordihydroaltenuenes A (**3**), colourless oil, $[\alpha]_D^{25}$ +113 (*c* 0.07, MeOH); UV(MeOH) λ_{max}(log ε) 211 (3.26), 269 (2.97), 307 (2.64) nm; ECD (0.24 mM, MeOH) λ_{max} (Δε) 207 (−15.34), 232 (+7.75), 246 (+0.43) and 271 (+6.89) nm; ^{1}H NMR (CD$_3$OD, 500 MHz) and ^{13}C NMR (CD$_3$OD, 125 MHz), Table 2; HRESIMS *m/z* 303.0844 [M + Na]$^+$ (calcd for $C_{14}H_{16}NaO_6$, 303.0839).

Isoochracinate A (**4** and **5**): yellow oil; $[\alpha]_D^{25}$ +1.6 (*c* 0.1, MeOH); UV (MeOH) λ_{max} (log ε) 207 (3.34), 233 (2.59), 300 (2.38) nm; ^{1}H NMR (CD$_3$OD, 500 MHz) and ^{13}C NMR (CD$_3$OD, 125 MHz), Table 2; HRESIMS *m/z* 223.0606 [M + H]$^+$ (calcd for $C_{11}H_{11}O_5$, 223.0607); (*S*)-isoochracinate A1 (**4**), $[\alpha]_D^{25}$ −21 (*c* 0.1, MeOH), ECD (0.45 mM, MeOH) λ_{max} (Δε) 219 (−1.25), 237 (+0.64), 247 (−1.12) and 301 (+0.70) nm; (*R*)-isoochracinate A2 (**5**), $[\alpha]_D^{25}$ +26 (*c* 0.1, MeOH), ECD (0.32 mM, MeOH) λ_{max} (Δε) 220 (+1.77), 235 (−0.76), 247 (+1.50) and 300 (−0.99) nm.

Alternariphent A (**6** and **7**): yellow powder; $[\alpha]_D^{25}$ −4.6 (*c* 0.1, MeOH); UV(MeOH) λ_{max}(log ε) 206 (2.37) nm; ^{1}H NMR (CD$_3$OD, 500 MHz) and ^{13}C NMR (CD$_3$OD, 125 MHz), Table 2; HRESIMS *m/z* 257.0787 [M + Na]$^+$ (calcd for $C_{13}H_{14}NaO_4$, 257.0784). (*S*)-Alternariphent A1 (**6**), $[\alpha]_D^{25}$ +3.1 (*c* 0.1, MeOH), ECD (0.64 mM, MeOH) λ_{max} (Δε) 211 (+0.47), 237 (−2.90), 259 (−1.46) and 334 (+0.59) nm; (*R*)-Alternariphent A2 (**7**), $[\alpha]_D^{25}$ −7.2 (*c* 0.1, MeOH), ECD (0.51 mM, MeOH) λ_{max} (Δε) 213 (−0.59), 233 (+4.53), 256 (+2.34) and 328 (−0.84) nm.

3.6. X-ray Crystal Structure Analysis

Crystallographic data for compound *nordihydroaltenuenes A* (**3**) were collected on a Rigaku XtaLAB PRO single-crystal diffractometer using Cu K*α* radiation. The structure of **3** was solved by direct methods (SHELXS 97), expanded using difference Fourier techniques, and refined by using the full-matrix least-squares calculation. The non-hydrogen atoms were refined anisotropically, and hydrogen atoms were fixed at calculated positions. Crystallographic data for the structure **3** has been deposited with the Cambridge Crystallographic Data Centre with the supplementary publication number CCDC-1847869. Copies of the data can be obtained free of charge from the CCDC at www. ccdc.cam.ac.uk.

Crystal data for **3**: Moiety formula: $C_{14}H_{16}O_6$ (M_W = 280.27), colourless block, crystal size = 0.2 × 0.1 × 0.1 mm^3, orthorhombic, space group P2$_1$2$_1$2$_1$; unit cell dimensions: *a* = 6.75000(10) Å, *b* = 8.05910(10) Å, *c* = 22.8555(3) Å, *V* = 1243.31(3) Å^3, *Z* = 4, ρ_{calcd} = 1.497 g cm^{-3}, *T* = 100.00(10) K, μ(Cu K*α*) = 0.995 mm^{-1}. A total of 5819 reflections were measured with 2434 independent reflections (R_{int} = 0.0209, R_{sigma} = 0.0222). Final *R* indices (I > 2σ (*I*)): R_1 = 0.0277, wR_2 = 0.0730. Final *R* indexes (all date): $R_1 = 0.0281$, $wR_2 = 0.0734$, Flack parameter = $-0.02(7)$. Largest diff. peak and hole = 0.15 and -0.19 eÅ^{-3}.

3.7. ECD Calculations

The theoretical calculations of new compound **1** was performed by using the density functional theory (DFT) as carried out in the Gaussian 03 [36]. Conformational analysis was initially conducted by using SYBYL-X 2.0 software. All ground-state geometries were optimized at the B3LYP/6-31G(d) level. TDDFT at B3LYP/6-31G(d) was employed to calculate the electronic excitation energies and rotational strengths in methanol [37,38]. Solvent effects of methanol solution were evaluated at the same DFT level by using the SCRF/PCM method [39].

3.8. Biological Assays

3.8.1. Antibacterial Activity Assay

Compounds **1–28** were tested for antibacterial activities against *Staphylococcus aureus* using the agar filter-paper diffusion method. Then, compounds which had an inhibition zone were evaluated in 96-well plates using a modified version of the broth microdilution method, and ampicillin was used as a positive control [40].

3.8.2. Antitumor Activity Assay

The in vitro cytotoxic activities against the three tumor cell lines (K562, SGC-7901 and BEL-7402) were assessed by the CCK-8 method, and the positive control was taxol [41].

4. Conclusions

Seven new structurally diverse polyketide derivatives (**1–7**), along with 21 known compounds (**8–28**), were isolated from cultures of the sponge-derived fungus, *Alternaria* sp. SCSIO41014. The structures and absolute configurations of these new compounds (**1–7**) were determined by spectroscopic analysis, X-ray single crystal diffraction, chiral separation, and comparison of ECD spectra to the calculations. Altertoxin VII (**1**) was the first example to possess a novel 4,8-dihydroxy substituted perylenequinone derivative, while the phenolic hydroxy groups always commonly substituted at C-4 and C-9. Compound **1** exhibited cytotoxic activities against the K562, SGC-7901, and BEL-7402 cell lines, with IC$_{50}$ values of 26.58 $\pm$ 0.80, 8.75 $\pm$ 0.13, and 13.11 $\pm$ 0.95 μg/mL, respectively. Compound **11** showed selectively cytotoxic activity against K562 with an IC$_{50}$ value of 19.67 $\pm$ 0.19 μg/mL. Compound **25** displayed moderate inhibitory activity against *Staphylococcus aureus*, with an MIC value of 31.25 μg/mL.

Supplementary Materials: The following materials are available online at http://www.mdpi.com/1660-3397/16/8/280/s1. The 16S rRNA gene sequences data of *Alternaria* sp. SCSIO41014, the 1D and 2D NMR spectra, HRESIMS for the new compounds **1–7**, and X-ray crystallographic files of compound **3** (in CIF format).

Author Contributions: X.P., J.W., and Y.L. contributed conception and design of the study. X.P. performed experiments, analyzed data and wrote the manuscript, X.L. did the isolation and identification of the fungus. All authors contributed to manuscript revision, read and approved the submitted version.

Funding: This research was funded by the National Natural Science Foundation of China (Nos. 21172230, 21772210, 21502204, 31270402, 41476135 and 41776169), the Strategic Priority Research Program of the Chinese Academy of Sciences (XDA11030403), the Science and Technology Project of Guangdong Province (No. 2016A020222010), and Pearl River S&T Nova Program of Guangzhou (No. 201710010136).

Conflicts of Interest: The authors declare no conflict of interest.

References

1. Chagas, F.O.; Dias, L.G.; Pupo, M.T. New perylenequinone derivatives from the endophytic fungus *Alternaria tenuissima* SS77. *Tetrahedron Lett.* **2016**, *57*, 3185–3189. [CrossRef]
2. Chowdhury, P.K.; Das, K.; Datta, A.; Liu, W.Z.; Zhang, H.Y.; Petrich, J.W. A comparison of the excited-state processes of nearly symmetrical perylene quinones: Hypocrellin A and hypomycin B. *J. Photoch. Photobiol. A* **2002**, *154*, 107–116. [CrossRef]
3. Mazzini, S.; Merlini, L.; Mondelli, R.; Scaglioni, L. Conformation and tautomerism of hypocrellins. Revised structure of shiraiachrome A. *J. Chem. Soc. Perk. Trans. 2* **2001**, 409–416. [CrossRef]
4. Zhang, N.D.; Zhang, C.Y.; Xiao, X.; Zhang, Q.Y.; Huang, B.K. New cytotoxic compounds of endophytic fungus *Alternaria* sp. isolated from *Broussonetia papyrifera* (L.) Vent. *Fitoterapia* **2016**, *110*, 173–180. [CrossRef] [PubMed]
5. Kjer, J.; Wray, V.; Edrada-Ebel, R.; Ebel, R.; Pretsch, A.; Lin, W.; Proksch, P. Xanalteric acids I and II and related phenolic compounds from an endophytic *Alternaria* sp. isolated from the mangrove plant *Sonneratia alba*. *J. Nat. Prod.* **2009**, *72*, 2053–2057. [CrossRef] [PubMed]
6. Gao, S.; Li, X.; Wang, B. Perylene derivatives produced by *Alternaria alternata*, An endophytic fungus isolated from *Laurencia* species. *Nat. Prod. Commun.* **2009**, *4*, 1477–1480. [PubMed]
7. Hradil, C.M.; Hallock, Y.F.; Clardy, J.; Kenfield, D.; Strobel, G. Phytotoxins from *Alternaria cassia*. *Phytochemistry* **1989**, *28*, 73–75. [CrossRef]
8. Bugni, T.S.; Ireland, C.M. Marine-derived fungi: A chemically and biologically diverse group of microorganisms. *Nat. Prod. Rep.* **2004**, *21*, 143–163. [CrossRef] [PubMed]
9. Wang, J.F.; Wei, X.Y.; Qin, X.C.; Lin, X.P.; Zhou, X.F.; Liao, S.R.; Yang, B.; Liu, J.; Tu, Z.C.; Liu, Y.H. Arthpyrones A-C, Pyridone alkaloids from a sponge-derived fungus *Arthrinium arundinis* ZSDS1-F3. *Org. Lett.* **2015**, *17*, 656–659. [CrossRef] [PubMed]
10. Pang, X.Y.; Lin, X.P.; Wang, J.F.; Liang, R.; Tian, Y.Q.; Salendra, L.; Luo, X.W.; Zhou, X.F.; Yang, B.; Tu, Z.C.; et al. Three new highly oxygenated sterols and one new dihydroisocoumarin from the marine sponge-derived fungus *Cladosporium* sp. SCSIO41007. *Steroids* **2018**, *129*, 41–46. [CrossRef] [PubMed]
11. Pang, X.Y.; Lin, X.P.; Tian, Y.Q.; Liang, R.; Wang, J.F.; Yang, B.; Zhou, X.F.; Kaliyaperumal, K.; Luo, X.W.; Tu, Z.C.; et al. Three new polyketides from the marine sponge-derived fungus *Trichoderma* sp. SCSIO41004. *Nat. Prod. Res.* **2018**, *32*, 105–111. [CrossRef] [PubMed]
12. Tian, Y.Q.; Lin, X.P.; Wang, Z.; Zhou, X.F.; Qin, X.C.; Kaliyaperumal, K.; Zhang, T.Y.; Tu, Z.C.; Liu, Y. Asteltoxins with antiviral activities from the marine sponge-derived fungus *Aspergillus* sp. SCSIO XWS02F40. *Molecules* **2016**, *21*, 34. [CrossRef] [PubMed]
13. Wang, J.F.; Lin, X.P.; Qin, C.; Liao, S.R.; Wan, J.T.; Zhang, T.Y.; Liu, J.; Fredimoses, M.; Chen, H.; Yang, B.; et al. Antimicrobial and antiviral sesquiterpenoids from sponge-associated fungus, *Aspergillus sydowii* ZSDS1-F6. *J. Antibiot.* **2014**, *67*, 581–583. [CrossRef] [PubMed]
14. Li, X.; Jing, P.Z.; Xu, N.L.; Meng, D.L.; Sha, Y. A new perylenequinone from the fruit bodies of *Bulgaria inquinans*. *J. Asian Nat. Prod. Res.* **2006**, *8*, 743–746.
15. Jiao, P.; Gloer, J.B.; Campbell, J.; Shearer, C.A. Altenuene derivatives from an unidentified freshwater fungus in the family Tubeufiaceae. *J. Nat. Prod.* **2006**, *69*, 612–615. [CrossRef] [PubMed]
16. Phaopongthai, J.; Wiyakrutta, S.; Meksuriyen, D.; Sriubolmas, N.; Suwanborirux, K. Azole-synergistic anti-candidal activity of altenusin, A biphenyl metabolite of the endophytic fungus *Alternaria alternata* isolated from *Terminalia chebula* Retz. *J. Microbiol.* **2013**, *51*, 821–828. [CrossRef] [PubMed]
17. Chou, T.H.; Chen, I.S.; Hwang, T.L.; Wang, T.C.; Lee, T.H.; Cheng, L.Y.; Chang, Y.C.; Cho, J.Y.; Chen, J.J. Phthalides from *Pittosporum illicioides* var. illicioides with inhibitory activity on superoxide generation and elastase release by neutrophils. *J. Nat. Prod.* **2008**, *71*, 1692–1695. [CrossRef] [PubMed]
18. Tianpanich, K.; Prachya, S.; Wiyakrutta, S.; Mahidol, C.; Ruchirawat, S.; Kittakoop, P. Radical scavenging and antioxidant activities of isocoumarins and a phthalide from the endophytic fungus *Colletotrichum* sp. *J. Nat. Prod.* **2011**, *74*, 79–81. [CrossRef] [PubMed]
19. Nakano, H.; Kumagai, N.; Matsuzaki, H.; Kabuto, C.; Hongo, H. Enantioselective addition of diethylzinc to aldehydes using 2-azanorbornylmethanols and 2-azanorbornylmethanethiol as acatalyst. *Tetrahedron Lett.* **1997**, *8*, 1391–1401. [CrossRef]

20. Takahashi, H.; Tsubuki, T.; Higashiyama, K. Highly diastereoselective reaction of chiral o-[2-(1,3-oxazolidinyl)] benzaldehydes with alkylmetallic reagents: Synthesis of chiral 3-substituted phthalides. *Chem. Pharm. Bull.* **1991**, *39*, 3136–3139. [CrossRef]

21. Liu, Y.; Yang, Q.; Xia, G.; Huang, H.; Li, H.; Ma, L.; Lu, Y.; He, L.; Xia, X.; She, Z. Polyketides with alpha-glucosidase inhibitory activity from a mangrove endophytic fungus, *Penicillium* sp. HN29-3B1. *J. Nat. Prod.* **2015**, *78*, 1816–1822. [CrossRef] [PubMed]

22. Stack, M.E.; Mazzola, E.P.; Page, S.W.; Pohlan, A.E. Mutagenic perylenequinone metabolites of *Alternaria Alternata*: alteroxins I, II, and III. *J. Nat. Prod.* **1986**, *49*, 866–871. [CrossRef] [PubMed]

23. Zheng, C.J.; Fu, X.M.; Zhang, X.L.; Kong, W.W.; Wang, C.Y. Bioactive perylene derivatives from a soft coral-derived fungus *Alternaria* sp. (ZJ-2008017). *Chem. Nat. Compd.* **2015**, *51*, 766–768. [CrossRef]

24. Zhao, D.L.; Wang, D.; Tian, X.Y.; Cao, F.; Li, Y.Q.; Zhang, C.S. Anti-phytopathogenic and cytotoxic activities of crude extracts and secondary metabolites of marine-derived fungi. *Mar. Drugs* **2018**, *16*, 36. [CrossRef] [PubMed]

25. Liu, Y.; Wu, Y.; Zhai, R.; Liu, Z.; Huang, X.; She, Z. Altenusin derivatives from mangrove endophytic fungus *Alternaria* sp. SK6YW3L. *RSC Adv.* **2016**, *6*, 72127–72132. [CrossRef]

26. Wang, Q.X.; Bao, L.; Yang, X.L.; Guo, H.; Yang, R.N.; Ren, B.; Zhang, L.X.; Dai, H.Q.; Guo, L.D.; Liu, H.W. Polyketides with antimicrobial activity from the solid culture of an endolichenic fungus *Ulocladium* sp. *Fitoterapia* **2012**, *83*, 209–214. [CrossRef] [PubMed]

27. Bradburn, N.; Coker, R.D.; Blunedn, G.; Turner, C.H.; Crabb, T.A. 5-epialtenuene and neoaltenuene, dibenzo-a-pyrones from *Alternaria alternata* cultured on rice. *Phtochemistry* **1994**, *35*, 665–669. [CrossRef]

28. Aly, A.H.; Edrada-Ebe, R.; Indriani, I.D.; Wray, V.; Muller, W.E.G.; Totzke, F.; Zirrgiebel, U.; Schachtele, C.; Kubbutat, M.H.G.; Lin, W.H.; et al. Cytotoxic metabolites from the fungal endophyte *Alternaria* sp. and their subsequent detection in its host plant polygonum senegalense. *J. Nat. Prod.* **2008**, *71*, 972–980. [CrossRef] [PubMed]

29. Wang, Y.; Yang, M.H.; Wang, X.B.; Li, T.X.; Kong, L.Y. Bioactive metabolites from the endophytic fungus *Alternaria alternata*. *Fitoterapia* **2014**, *99*, 153–158. [CrossRef] [PubMed]

30. Tian, J.; Fu, L.; Zhang, Z.; Dong, X.; Xu, D.; Mao, Z.; Liu, Y.; Lai, D.; Zhou, L. Dibenzo-alpha-pyrones from the endophytic fungus *Alternaria* sp. Samif01: Isolation, Structure elucidation, And their antibacterial and antioxidant activities. *Nat. Prod. Res.* **2016**, *31*, 387–396. [CrossRef] [PubMed]

31. Wang, Y.; Wang, L.; Zhuang, Y.; Kong, F.; Zhang, C.; Zhu, W. Phenolic polyketides from the co-cultivation of marine-derived *Penicillium* sp. WC-29-5 and *Streptomyces fradiae* 007. *Mar. Drugs* **2014**, *12*, 2079–2088. [CrossRef] [PubMed]

32. Kellogg, J.J.; Todd, D.A.; Egan, J.M.; Raja, H.A.; Oberlies, N.H.; Kvalheim, O.M.; Cech, N.B. Biochemometrics for natural products research: comparison of data analysis approaches and application to identification of bioactive compounds. *J. Nat. Prod.* **2016**, *79*, 376–386. [CrossRef] [PubMed]

33. De Souza, G.D.; Mithofer, A.; Daolio, C.; Schneider, B.; Rodrigues-Filho, E. Identification of *Alternaria alternata* mycotoxins by LC-SPE-NMR and their cytotoxic effects to soybean (*Glycine max*) cell suspension culture. *Molecules* **2013**, *18*, 2528–2538. [CrossRef] [PubMed]

34. Kim, N.; Sohn, M.J.; Koshino, H.; Kim, E.H.; Kim, W.G. Verrulactone C with an unprecedented dispiro skeleton, A new inhibitor of *Staphylococcus aureus* enoyl-ACP reductase, from *Penicillium verruculosum* F375. *Bioorg. Med. Chem. Lett.* **2014**, *24*, 83–86. [CrossRef] [PubMed]

35. Kashiwada, Y.; Nonaka, G.I.; Nishioka, I. Chromone glucosides from *Rhubarb*. *Phtochemistry* **1990**, *29*, 1007–1009. [CrossRef]

36. Frisch, M.J.; Trucks, G.W.; Schlegel, H.B.; Scuseria, G.E.; Robb, M.A.; Cheeseman, J.R.; Montgomery, J.A.; Vreven, T., Jr.; Kudin, K.N.; Burant, J.C.; et al. *Gaussian 03, Revision E.01*; Gaussian, Inc.: Wallingford, CT, USA, 2004; Available online: http://gaussian.com/g03citation/ (accessed on 8 August 2018).

37. Tomasi, J.; Persico, M. Molecular interactions in solution: An overview of methods based on continuous distributions of the solvent. *Chem. Rev.* **1994**, *94*, 2027–2094. [CrossRef]

38. Cammi, R.; Tomasi, J. Remarks on the use of the apparent surface charges (ASC) methods in solvation problems: Iterative versus Matrix-Inversion procedures and the renormalization of the apparent charges. *J. Comput. Chem.* **1995**, *16*, 1449–1458. [CrossRef]

39. Gross, E.K.U.; Dobson, J.F.; Petersilka, M. Density functional theory of time-dependent phenomena. *Top. Curr. Chem.* **1996**, *181*, 81–172.
40. Wang, J.F.; Cong, Z.W.; Huang, X.L.; Hou, C.X.; Chen, W.B.; Tu, Z.C.; Huang, D.Y.; Liu, Y.H. Soliseptide A, A cyclic hexapeptide possessing piperazic acid groups from *Streptomyces solisilvae* HNM30702. *Org. Lett.* **2018**, *20*, 1371–1374. [CrossRef] [PubMed]
41. Wang, J.F.; Wei, X.Y.; Qin, X.C.; Tian, X.P.; Liao, L.; Li, K.M.; Zhou, X.F.; Yang, X.W.; Wang, F.Z.; Zhang, T.Y.; et al. Antiviral merosesquiterpenoids produced by the antarctic fungus *Aspergillus ochraceopetaliformis* SCSIO 05702. *J. Nat. Prod.* **2016**, *79*, 59–65. [CrossRef] [PubMed]

 *marine drugs*

Article

Two Novel Aspochalasins from the Gut Fungus *Aspergillus* sp. Z4

Xinyang Li, Wanjing Ding, Pinmei Wang and Jinzhong Xu *

Institute of Marine Biology, Ocean College, Zhejiang University, Zhoushan 316021, China;
lxinyang2014@126.com (X.L.); wading@zju.edu.cn (W.D.); wangpinmei@zju.edu.cn (P.W.)
* Correspondence: xujinzhong@zju.edu.cn; Tel.: +86-158-581-680-18

Received: 2 September 2018; Accepted: 18 September 2018; Published: 20 September 2018

Abstract: Two novel aspochalasins, tricochalasin A (**1**) and aspochalasin A2 (**2**), along with three known compounds (**3–5**) have been isolated from the different culture broth of *Aspergillus* sp., which was found in the gut of a marine isopod *Ligia oceanica*. Compound **1** contains a rare 5/6/6 tricyclic ring fused with the aspochalasin skeleton. The structures were determined on the basis of electrospray ionisation mass spectroscopy (ESIMS), nuclear magnetic resonance (NMR) spectral data, and the absolute configurations were further confirmed by modified Mosher's method. Cytotoxicity against the prostate cancer PC3 cell line were assayed by the MTT method. Compound **3** showed strong activity while the remaining compounds showed weak activity.

Keywords: aspochalasin; tricyclic fused; gut fungus; cytotoxicity

1. Introduction

Aspochalasins constitute a subgroup within the small group of cytochalasans, which are fungal secondary metabolites known for varieties of biological activities [1]. These include cytotoxic [2–4], anti human immunodeficiency virus (HIV) [5], immunomodulatory [6], and nematicidal activity [7]. So far, more than 200 cytochalasan analogues have been reported [1]. Structurally, this group of compounds contains one isoindole unit fused with one macrocyclic ring. Isotope-labeling experiments have revealed that cytochalasans originate from an acetyl- and methionine-derived polyketide chain and the attachment of an amino acid precursors such as Leu, Phe, Ala and Trp [8,9]. With diverse oxygenated regions in the macrocyclic ring, there are several unusual analogues among these known compounds including chaetochalasin A [10], aspergillin PZ [11], spicochalasin A [12], epicochalasines A and B [13]. In our ongoing search for new bioactive metabolites of marine fungi, some new compounds have been purified from the marine-derived fungus Z4 [14–17], and one strain isolated from the gut of the marine isopod *Ligia oceanica*. In order to find more novel natural products from this fungus, we employed the OSMAC (one strain, many compounds) approach by varying the culture conditions of Z4. Two new cytochalasans, aspochalasins **1** and **2**, in addition to three known cytochalasans, aspochalasins **3–5** (Figure 1) were purified when cultured in media 2216E and rice. Herein we present the isolation, structure elucidation, and cytotoxic activity of these aspochalasins.

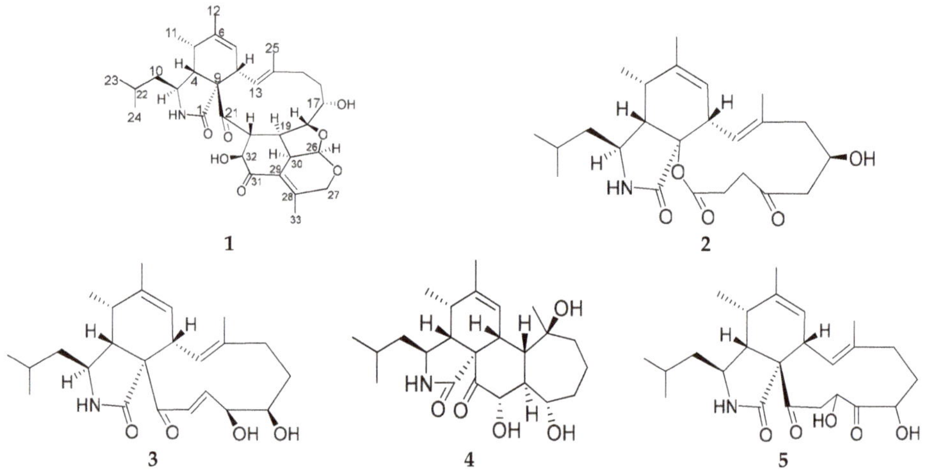

Figure 1. Structures of compounds **1–5**.

2. Results and Discussion

Compound **1** was purified as a colorless solid. The molecular formula $C_{32}H_{43}NO_7$ with 12 degrees of unsaturation was established by the positive mode quasi-molecular ion peaks at *m/z* 554.3107 for $[M + H]^+$ (calcd. 554.3073 for $C_{32}H_{44}NO_7$) and *m/z* 576.2930 for $[M + Na]^+$ (calcd. 576.2893 for $C_{32}H_{43}NO_7Na$) combined with 1D nuclear magnetic resonance (NMR) data. The ^1H NMR spectrum recorded in $CDCl_3$ together with heteronuclear single quantum coherence (HSQC) spectra revealed six methyl groups [δ_H 0.93 (3H, d, *J* = 6.4 Hz), 0.96 (3H, d, *J* = 6.4 Hz), 1.18 (3H, d, *J* = 7.1 Hz), 1.57 (3H, s), 1.77 (3H, s), 1.85 (3H, s)], two olefinic protons [δ_H 5.45 (1H, br s), δ_H 5.70 (1H, d, *J* = 11.4 Hz)], three oxygenated methine hydrogen atoms [δ_H 4.35 (1H, d, *J* = 6.29 Hz), δ_H 4.42 (1H, dd, *J* = 3.6, 11.1 Hz), δ_H 3.86 (1H, d, *J* = 11.3 Hz)], one oxygenated methylene proton [δ_H 4.79 (2H, m)], one proton of acetal group [δ_H 5.91 (1H, d, *J* = 3.9 Hz)] and other 18 aliphatic protons. The ^{13}C NMR and distortionless enhancement by polarization transfer (DEPT) spectra of **1** indicated 32 carbon resonances ascribed to two ketone carbonyl, one amide carbonyl, six olefinic carbons, five oxygenated carbons including one acetal carbon, 12 aliphatic carbon atoms and six methyl carbons (Table 1). These features characteristically revealed the structure of **1** to be an aspochalasin skeleton. The excess number of carbon atoms when compared to the previously isolated and reported aspochalasin derivatives [18,19], indicated that **1** was an unusual one.

Table 1. Nuclear magnetic resonance (NMR) spectroscopic data of compounds **1** and **2** in $CDCl_3$.

Pos.	1 δ_C [a], Type	1 δ_H (*J* in Hz) [b]	2 δ_C [a], Type	2 δ_H (*J* in Hz) [b]
1	176.4, C		172.4, C	
2		6.10, brs		5.96, brs
3	52.9, CH	2.97, m	52.4, CH	2.99, m
4	53.9, CH	2.81, t (4.4)	52.3, CH	2.56, dd (8.5,7.1)
5	35.9, CH	2.67, brs	34.8, CH	2.80, brs
6	140.6, C		140.9, C	
7	125.6, CH	5.45, brs	123.5, CH	5.29, brs
8	42.9, CH	3.61, d (12.1)	41.1, CH	3.36, d (10.3)
9	68.9, C		86.5, C	
10	47.2, CH_2	1.49, m; 1.25, m	46.8, CH_2	1.83, m; 1.23, m
11	14.4, CH_3	1.18, d (7.1)	14.3, CH_3	1.19, d (7.3)
12	20.5, CH_3	1.77, s	20.1, CH_3	1.77, s
13	125.1, CH	5.70, d (11.4)	123.8, CH	5.94, d (10.3)
14	135.9, C		138.2, C	

Table 1. *Cont.*

Pos.	1		2	
	δ_C [a], Type	δ_H (*J* in Hz) [b]	δ_C [a], Type	δ_H (*J* in Hz) [b]
15	36.9, CH$_2$	2.07, m	33.1, CH$_2$	2.27, m; 2.14, m
16	29.3, CH	1.50, m; 1.30, m	75.1, CH	4.42, brs
17	68.6, CH	4.35, d (6.3)	32.3, CH$_2$	2.04, m; 2.28, m
18	88.8, CH	3.86, d (11.3)	208.9, C	
19	44.0, CH	2.25, m	29.0, CH$_2$	2.32, m; 3.02, m
20	47.7, CH	3.51, dd (4.9, 11.0)	34.1, CH$_2$	2.62, m; 2.44, m
21	214.2, C		171.9, C	
22	26.0, CH	1.53, m	25.6, CH	1.59, m
23	21.2, CH$_3$	0.93, d (6.4)	21.1, CH$_3$	0.92, d (6.5)
24	24.2, CH$_3$	0.96, d (6.4)	23.8, CH$_3$	0.94, d (6.5)
25	16.1, CH$_3$	1.57, s	17.7, CH$_3$	1.56, s
26	109.4, CH	5.91, d (3.9)		
27	73.1, CH$_2$	4.79, m		
28	158.1, C			
29	127.6, C			
30	51.4, CH	3.59, m		
31	197.0, C			
32	77.3, CH	4.42, dd (3.6, 11.1)		
33	14.0, CH$_3$	1.85, s		

[a] Recorded at 500 MHz in CDCl$_3$. [b] Recorded at 125 MHz in CDCl$_3$.

The planar structure of **1** was elucidated by 2D NMR spectrum. The ^{1}H-^{1}H correlation spectroscopy (COSY) cross peaks (Figure S8 in Supplementary Materials) of Me-23 (δ_H 0.93)/Me-24 (δ_H 0.96)/H-22 (δ_H 1.53)/H-10 (δ_H 1.49, 1.25)/H-3 (δ_H 2.97)/H-4 (δ_H 2.81)/H-5 (δ_H 2.67)/Me-11 (δ_H 1.18), H-7 (δ_H 5.45)/H-8 (δ_H 3.61)/H-13 (δ_H 5.70), H-15 (δ_H 2.07)/H-16 (δ_H 1.50, 1.30)/H-17 (δ_H 4.35) and heteronuclear multiple bond correlation (HMBC) correlations (Figure S9) from Me-12 (δ_H 1.77) to C-5 (δ_C 35.9), C-6 (δ_C 140.6) and C-7 (δ_C 125.6), from Me-25 (δ_H 1.57) to C-13 (δ_C 125.1), C-14 (δ_C 135.9) and C-15 (δ_C 36.9) established unit A, possessing a (2-methylpropyl) isoindolone moiety, which had two positions vacant to be linked, R1 and R2 (Figure 2). 11 carbon signals remained. The ^{1}H-^{1}H COSY cross peaks of H-26 (δ_H 5.91)/H-30 (δ_H 3.59)/H-19 (δ_H 2.25)/H-18 (δ_H 3.86) and H-19/H-20 (δ_H 3.51)/H-32 (δ_H 4.42)/OH-32 (δ_H 4.58), together with HMBC interactions from Me-33 (δ_H 1.85) to C-29 (δ_C 127.6), C-28 (δ_C 158.1) and C-31 (δ_C 197.0), from H-20 and OH-32 to C-31 were preliminarily attributed to unit B (Figure 2). Furthermore, the HMBC correlations from H-19 and H-20 to C-21 (δ_C 214.2), from H-18 to C-16 (δ_C 29.3), C-17 (δ_C 68.6) and C-19 connected R1 with R4, and R2 with R3 respectively. To satisfy the unsaturation, R5 and R6 were linked. Thus, the gross planar structure of **1** was established.

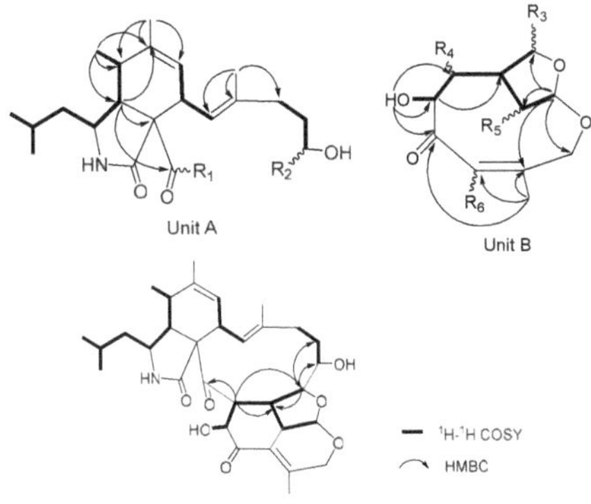

Figure 2. Partial structures of **1** based on ^{1}H-^{1}H correlation spectroscopy (COSY) and heteronuclear multiple bond correlation (HMBC) spectra.

The relative stereochemistry of **1** was determined with the help of ^{1}H NMR coupling constants, nuclear Overhauser effect spectroscopy (NOESY) experiments, and comparison with those of reported aspochalasins. The NOESY cross peaks (Figure S10) among H-5, H-4 and H-8 demonstrated the relative configurations of the isoindolone moiety in accordance with those of reported cytochalasans [20,21]. For the macrocyclic part, the NOESY correlations of H-17, H-18 and H-20 suggested they were cofacial. The double peak of H-18 with big coupling constants between H-19 reflected the *trans*-orientation of these two atoms. For the penta-heterocycle moiety, both H-19 and H-30 were located at the joint of three cycles, which elucidated the axial bond in these two atoms. Additionally, the NOESY correlations between H-19, H-30, H-26 and H-32 established that these moieties had the same orientation (Figure 3). The modified Mosher's method using $(S)/(R)$-α-methoxy-α-(trifluoromethyl) phenylacetyl (MPTA)-Cl was applied to assign the absolute configuration of **1**. The positive and negative value disposition ($\Delta\delta^{S-R}$) of the Mosher's ester derivatives (**1a** and **1b**) established the absolute configuration of C-17 as *S* (Figure 4). It is noteworthy that in all natural cytochalasans, so far, the stereochemistry of perhydroisoindol-1-one moiety is the same [4,21] which assigned the absolute configurations for C-3, C-4, C-5,C-8 and C-9 as 3*S*, 4*R*, 5*S*, 8*S*, 9*S*, respectively. Therefore, the complete absolute stereochemistry of **1** could be assigned as 3*S*, 4*R*, 5*S*, 8*S*, 9*S*, 17*S*, 18*R*, 19*R*, 20*R*, 26*R*, 30*S*, 32*S* and named as Tricochalasin A.

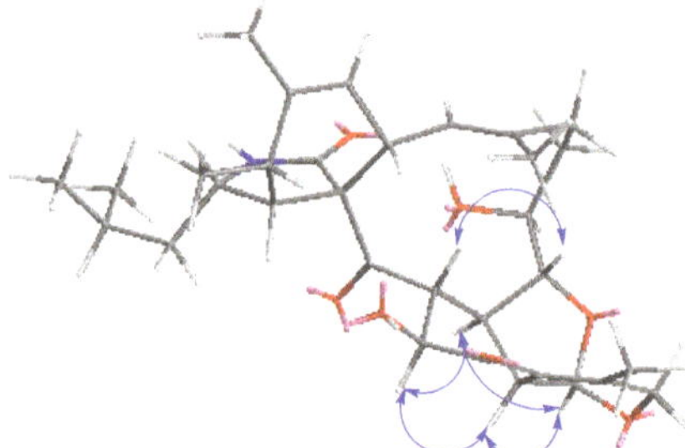

Figure 3. Key nuclear Overhauser effect spectroscopy (NOESY) correlations of **1**.

Figure 4. $\Delta\delta$ values (in ppm) = δ_S-δ_R obtained for (S)- and (R)-α-methoxy-α-(trifluoromethyl) phenylacetyl (MPTA)-Cl esters **1a** and **1b**.

Compound **2** was isolated as a white solid. Its molecular formula was deduced to be $C_{24}H_{35}NO_5$ (8 degrees of unsaturation) by positive-mode high-resolution electrospray ionization mass spectrometry (HR-ESIMS) ion peaks at m/z 418.2589 [M + H]$^+$ (calcd. 418.2549 for $C_{24}H_{36}NO_5$) and m/z 440.2406 [M + Na]$^+$ (calcd. 440.2368 for $C_{24}H_{35}NO_5Na$), combined with 1D NMR data. Analysis of the ^{1}H NMR and ^{13}C NMR spectrum data recorded in CDCl$_3$ revealed the presence of (2-methylpropyl) isoindolone moiety similar to those found in **1**. ^{13}C NMR and DEPT spectra showed 24 carbon resonances including one ketone carbonyl (δ_C 208.9), one ester/lactone carbonyl (δ_C 171.9), one amide carbonyl (δ_C 172.4), four olefinic carbon (δ_C 123.5, 123.8, 138.2 and 140.9), five aliphatic methylene carbons (δ_C 46.8, 33.1, 32.3, 29.0 and 34.1), six aliphatic methine carbons (δ_C 25.6, 52.3, 52.4, 34.8, 41.1 and 75.1), five methyl signals (δ_C 21.1, 23.8, 14.3, 20.1 and 17.7) and one quaternary carbon (δ_C 86.5) (Table 1). These features characteristically suggested **2** belongs to the same structural family as **1**. Comparison with reported aspochalasin derivatives indicated that **2** was similar to aspochalasin A1 [22]. The planar structure of **2** was elucidated by ^{1}H-^{1}H COSY and ^{1}H-^{13}C HMBC experiments (Figure 5). ^{1}H-^{1}H COSY cross peaks (Figure S20) of the H-15 (δ_H 2.14, 2.27)/H-16 (δ_H 4.42)/H-17 (δ_H 2.04, 2.28) established the position of the hydroxyl group at C-16 (δ_C 75.1) and the H-19 (δ_H 2.32, 3.02)/H-20 (δ_H 2.44, 2.62) pairs located the ketone carbonyl group at C-18 (δ_C 208.9). The key HMBC correlations (Figure S21) from H-17 to C-16 and C-18, from H-15 to C-16 and C-14 (δ_C 138.2) and from H-19, H-20 to C-18 and C-21 (δ_C 171.9) also demonstrated the structural features.

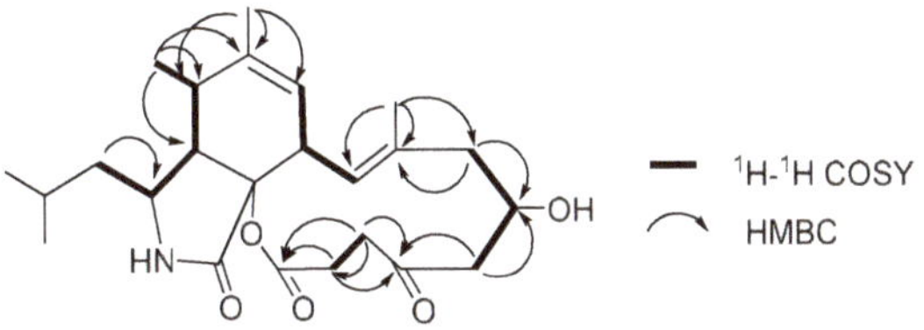

Figure 5. ^{1}H-^{1}H COSY and key HMBC correlations of **2**.

The absolute configuration of C-16 in **2** was established by the convenient Mosher's ester. The difference in chemical shift values of the easters **2a** and **2b** was calculated to assign the absolute configuration at C-16 as *R* (Figure 6). Thus, the absolute configuration of **2** was deduced as 3*S*, 4*R*, 5*S*, 8*S*, 9*S*, 16*R* and named as aspochalasin A2.

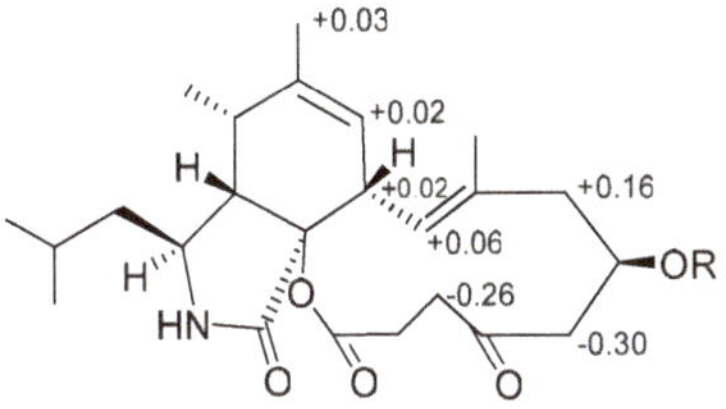

Figure 6. $\Delta\delta$ values (in ppm) = δ_S-δ_R obtained for (*S*)- and (*R*)-MPTA esters **2a** and **2b**.

NMR data of compounds **3–5** was in full agreement with the previously reported values for aspochalasins D (**3**) [23], aspergilluchalasin (**4**) [24] and aspochalasins T (**5**) [25].

All compounds were tested for their in vitro cytotoxicity against the prostate cancer PC3 cell lines by the MTT method, using doxorubicin (ADR) as positive control. As shown in Table 2, compound **3** showed strong activity against PC3 cell line, while others showed weak activities against it.

Table 2. Growth inhibition of **1–5** against prostate cancer PC3 cell line.

Compounds	PC3 Cell Line (IC$_{50}$ in μM)
1	>36
2	>40
3	11.14
4	>40
5	>40
ADR	5.09

3. Materials and Methods

3.1. General Experimental Procedures

Optical rotations were measured on a Jasco P-1010 polarimeter. The ultraviolet (UV) absorption spectra were measured in MeOH on a METASH UV-8000 spectrophotometer. The infrared (IR) spectra were recorded using a Bruker Vector 22 spectrometer (film) and an Avatar 370 Fourier-transform–infrared (FT–IR) spectrometer (KBr disk). NMR spectra were recorded in Chloroform-*d* (ALDRICH, St. Louis, MO, USA) with tetramethylsilane as an internal standard or Pyridine-d_5 (ALDRICH, St. Louis, MO, USA), using a Bruker AVANCE-III 500 MHz NMR spectrometer (Brucker, Ettlingen, Germany). HR-ESIMS data were obtained on an Agilent 6224 TOF LC/MS (Agilent, Santa Clara, CA, USA) (positive mode; infusion rate: 0.6 mL/min; capillary voltages: 4000 V; nebulizer: 50 psig; drying gas: 10mL/min; gas temperature: 325 °C).

Column chromatography (CC) was performed with silica (100–200 and 200–300 mesh, Qingdao Haiyang Chemical Co., Ltd., Qingdao, China).

3.2. Fungal Material

The fungal colony (Z4) was isolated from marine isopod *Ligia oceanica* which was collected in seaside of Dinghai in Zhoushan, Zhejiang Province of China in December 2011. The fungal Z4 was determined as *Aspergillus* sp. by 18S rDNA analysis and preserved in China Center for Type Culture Collection (CCTCC No.M2013631).

3.3. Fermentation, Extraction, and Isolation of Strain Z4

Two different media were used: A—500 mL Erlenmeyer-flasks (400 × 200 mL, a total of 80 L) each containing 200 mL of 2216E liquid media (QingDao Hopebio-Technology Co., Ltd., Qingdao, China) and B—40 500 mL Erlenmeyer-flasks each containing 40 g of rice with 45 mL 3% artificial seawater. Both medium were incubated at 28 °C for 30 days in static condition.

Medium A was filtered and extracted with equivoluminal EtOAc for 3 times to yield 5 g of extract, which was fractioned by silica gel column chromatography (CC) eluted in gradient from petroleum ether-EtOAc (100:1–2:1) to CH_2Cl_2-MeOH (50:1–0:100). 8 fractions were collected based on thin layer chromatography (TLC) analysis. Fr. 5 (2.5 g) was fractionated by CC over silica gel using a gradient of CH_2Cl_2-MeOH (100:1–0:100) as a mobile phase to provide eight subfractions (Frs. 5-1 to 5-8). Fr. 5-4 (0.5 g) was further purified by CC over silica gel using a gradient of cyclohexane(C)-E(3:1–1:1) to yield nine subfractions (Frs. 5-4-1 to 5-4-9). Fr. 5-4-5 (90 mg) were purified by semi-preparative octadecyl-silica high-performance liquid chromatography (ODS-HPLC) (COSMOSIL PACKED COLUMN, 5C18-MS-II column, 10ID × 250 mm, 3 mL/min, 54% acetonitrile in H_2O) to obtain compound **2** (3 mg). Fr. 5-4-9 (37 mg) were purified by semi-preparative ODS-HPLC (3 mL/min, 52% MeOH in H_2O) to obtain compound **5** (17 mg). Each flask of B medium was extracted 3 times with EtOAc (60 mL) yielding 39 g of extract. CC over silica gel using a gradient of CH_2Cl_2-MeOH (100:1–0:100) as the eluents to afforded seven fractions based on TLC analysis. Fr. 3 (900 mg) was then purified by CC over silica gel with a CH_2Cl_2-MeOH gradient from 100:1–10:1 to afford 10 subfractions (Frs. 3-1 to 3-10), among which, Fr. 3-6 (28 mg) was further purified by semi-preparative ODS-HPLC (4 mL/min, 75% MeOH in H_2O) to obtain compounds **1** (3 mg). Fr. 6 (4 g) was purified by CC over silica gel with a CH_2Cl_2-MeOH gradient from 50:1–5:1 to afford six subfractions (Frs. 6-1 to 6-6). Fr. 6-2 (76 mg) was purified by semi-preparative ODS-HPLC (4 mL/min, 75% MeOH in H_2O) to obtain compounds **4** (13 mg). Fr. 6-3 (28 mg) was purified by semi-preparative ODS-HPLC (4 mL/min, 80% MeOH in H_2O) to yield compound **3** (1 mg).

Tricochalasin A (**1**): colorless solid; $[\alpha]_D^{25}$ −2.39 (c 0.67, MeOH); UV (MeOH) λ_{max} (log ε) 204 (3.4), 245 (1.6) nm; IR(film) ν_{max} 3391, 2931, 1753, 1681, 1452, 1380, 1189, 1095 cm^{-1}; ^{1}H and ^{13}C data see Table 1; HR-ESIMS *m*/*z* 554.3107 [M + H]$^+$, calcd. 554.3073 for $C_{32}H_{44}NO_7$.

Aspochalasin A2 (**2**): colorless solid; $[\alpha]_D^{25}$ 8.08 (c 0.98, CHCl$_3$); UV (MeOH) λ_{max} (log ε) 202 (3.6) nm; IR(KBr) ν_{max} 3454, 3273, 2924, 2852, 1709, 1435, 1384, 1324, 1263, 1142 cm^{-1}; ^{1}H and ^{13}C data see Table 1; HR-ESIMS *m/z* 418.2589 [M + H]$^+$, calcd. 418.2549 for C$_{24}$H$_{36}$NO$_5$.

Aspochalasins D (**3**): colorless solid; UV (MeOH) λ_{max} (log ε) 204 (3.1) nm.

Aspergillulactone (**4**): colorless solid; UV (MeOH) λ_{max} (log ε) 204 (3.6), 245 (1.3) nm.

Aspochalasins T (**5**): colorless solid; UV (MeOH) λ_{max} (log ε) 205 (3.7) nm.

3.4. Preparation of MTPA Esters

Compound **1** was transferred into two clean NMR tubes (0.5 mg in each tube) and dried completely under vacuum. Deuterated pyridine (400 µL) was added to dissolve the sample and (R)-MTPA-Cl/(S)-MTPA-Cl (8µL) were quickly added into the tubes, respectively. All contents were mixed thoroughly by shaking the tubes carefully. The reaction was performed at room temperature for 4h to obtained the S-MTPA and the R-MTPA ester (**1a** and **1b**), respectively. The chemical shift differences ($\Delta\delta = \delta_S - \delta_R$) calculated from the ^{1}H NMR spectra of the two diastereomeric esters **1a** and **1b**, obtained without purification, enabled us to determine the absolute configuration of C-17 as *S*.

Similarly, the S-MTPA and R-MTPA ester (**2a** and **2b**) of compound **2** was obtained. ^{1}H NMR data of **2a** and **2b** were run without further purification and the chemical shift differences ($\Delta\delta = \delta_S - \delta_R$) enabled us to determine the absolute configuration of C-16 as *R*.

1a: ^{1}H NMR (500 MHz, pyridine-d_5): δ 6.68 (1H, H-13), δ 1.37 (1H, H-16a), δ 7.86 (1H, H-17), δ 4.51 (1H, H-18), δ 2.77 (1H, H-19), δ 4.95 (1H, H-27a), δ 4.03 (1H, H-30), δ 4.32 (1H, H-32).

1b: ^{1}H NMR (500 MHz, pyridine-d_5): δ 6.72 (1H, H-13), δ 1.42 (1H, H-16a), δ 7.92 (1H, H-17), δ 4.09 (1H, H-18), δ 2.71 (1H, H-19), δ 4.93 (1H, H-27a), δ 4.01 (1H, H-30), δ 4.25 (1H, H-32).

2a: ^{1}H NMR (500 MHz, pyridine-d_5): δ 5.42 (1H, H-7), δ 3.26 (1H, H-8), δ 1.79 (3H, H-12), δ 6.61 (1H, H-13), δ 2.35 (1H, H-15a), δ 2.25 (1H, H-17a), δ 2.14 (1H, H-19a).

2b: ^{1}H NMR (500 MHz, pyridine-d_5): δ 5.40 (1H, H-7), δ 3.24 (1H, H-8), δ 1.76 (3H, H-12), δ 6.55 (1H, H-13), δ 2.19 (1H, H-15a), δ 2.55 (1H, H-17a), δ 2.40 (1H, H-19a).

3.5. Cytotoxicity Bioassays

The cytotoxicity was measured by the MTT assay against the prostate cancer PC3 cell line. Tumor cell lines were seeded in 96-well plates (4×10^3 per well in 100 µL). After 24 h of incubation, the cells were treated with gradient concentrations (100 µM, 50 µM, 25 µM, 12.5 µM, 6.25 µM, 3.125 µM) for another 72 h. Afterwards, MTT solution (5.0 mg/mL in RPMI-1640 media, Sigma, St. Louis, MO, USA) was added (20 µL/well) and the plates were incubated for another 4 h at 37 °C. The compounds were dissolved in DMSO and cell growth inhibition assay was performed as reported previously [26]. The growth inhibitory ability of the compounds were calculated and expressed using the IC$_{50}$ value by dose-effect analysis software. Doxorubicin (ADR) was used as a positive control.

4. Conclusions

A chemical investigation was carried out on the marine fungus Z4, which resulted in the isolation of two novel aspochalasins (compounds **1** and **2**) along with three known aspochalasins (compounds **3–5**). The planar structures were determined and compound **1** contains a rare 5/6/6 tricyclic ring fused with the aspochalasin skeleton. Furthermore, the absolute configurations were established by modified Mosher's method. Cytotoxicity against the prostate cancer PC3 cell line were assayed by MTT method. Compound **3** showed strong activity while others showed weak activity.

Supplementary Materials: The following are available online at http://www.mdpi.com/1660-3397/16/10/343/s1, Figure S1: IR spectrum of **1**; Figures S2 and S3: HR-ESI-MS spectrum of **1**; Figures S4–S10: ^{1}H, ^{13}C, DEPT,

HSQC, COSY, HMBC, NOESY NMR data of **1** in CDCl$_3$; Figures S11 and S12: ^{1}H NMR data of **1a** and **1b** in pyridine-d_5; Figure S13: IR spectrum of **2**; Figures S14 and S15: HR-ESI-MS spectrum of **2**; Figures S16–S22: ^{1}H, ^{13}C, DEPT, HSQC, COSY, HMBC, NOESY NMR data of **2** in CDCl$_3$; Figures S23 and S24: ^{1}H NMR data of **2a** and **2b** in pyridine-d_5; Figures S25 and S26: ^{1}H and ^{13}C NMR data of **3** in CDCl$_3$; Figures S27 and S28: ^{1}H and ^{13}C NMR data of **4** in CDCl$_3$; Figures S29 and S30: ^{1}H and ^{13}C NMR data of **5** in CDCl$_3$.

Author Contributions: J.X. designed the experiment. X.L. performed the experiment, determined the structures and prepare the manuscript. W.D. tested the biological activities. P.W. isolated and identified the fungus.

Funding: This work was supported by the Natural Science Foundation of China (NSFC No. 41406141).

Acknowledgments: We thank Jianyang Pan (Pharmaceutical Informatics Institute, Zhejiang University) for performing NMR spectroscopy for structure elucidation.

Conflicts of Interest: The authors declare no conflict of interest.

References

1. Scherlach, K.; Boettger, D.; Remme, N.; Hertweck, C. The chemistry and biology of cytochalasans. *Nat. Prod. Rep.* **2010**, *27*, 869–886. [CrossRef] [PubMed]

2. Zhang, D.; Ge, H.; Xie, D.; Chen, R.; Zou, J.H.; Tao, X.; Dai, J. Periconiasins A–C, new cytotoxic cytochalasans with an unprecedented 9/6/5 tricyclic ring system from endophytic fungus *periconia* sp. *Org. Lett.* **2013**, *15*, 1674–1677. [CrossRef] [PubMed]

3. Knudsen, P.B.; Hanna, B.; Ohl, S.; Sellner, L.; Zenz, T.; Dohner, H.; Stilgenbauer, S.; Larsen, T.O.; Lichter, P.; Seiffert, M. Chaetoglobosin A preferentially induces apoptosis in chronic lymphocytic leukemia cells by targeting the cytoskeleton. *Leukemia* **2014**, *28*, 1289–1298. [CrossRef] [PubMed]

4. Zhou, G.X.; Wijeratne, E.M.K.; Bigelow, D.; Pierson, L.S.; Vanetten, H.D.; Guantilaka, A.A.L. Aspochalasins I, J, and K: Three new cytotoxic cytochalasans of *Aspergillus flavipes* from the rhizosphere of *Ericameria laricifolia* of the sonoran desert. *J. Nat. Prod.* **2004**, *67*, 328–332. [CrossRef] [PubMed]

5. Rochfort, S.; Ford, J.; Ovenden, S.; Wan, S.S.; George, S.; Wildman, H.; Tait, R.M.; Meurer-Grimes, B.; Cox, S.; Coates, J.; et al. A novel aspochalasin with HIV-1 integrase inhibitory activity from *Aspergillus flavipes*. *J. Antibiot.* **2005**, *58*, 279–283. [CrossRef] [PubMed]

6. Hua, C.Y.; Yang, Y.H.; Sun, L.; Dou, H.; Tan, R.X.; Hou, Y.Y. Chaetoglobosin F, a small molecule compound, possesses immunomodulatory properties on bone marrow-derived dendritic cells via TLR9 signaling pathway. *Immunobiology* **2013**, *218*, 292–302. [CrossRef] [PubMed]

7. Hu, Y.; Zhang, W.P.; Zhang, P.; Ruan, W.B.; Zhu, X.D. Nematicidal Activity of Chaetoglobosin A Poduced by *Chaetomium globosum* NK102 against Meloidogyne incognita. *J. Agric. Food Chem.* **2013**, *61*, 41–46. [CrossRef] [PubMed]

8. Qiao, K.J.; Chooi, Y.H.; Tang, Y. Identification and engineering of the cytochalasin gene cluster from Aspergillus clavatus NRRL 1. *Metab. Eng.* **2011**, *13*, 723–732. [CrossRef] [PubMed]

9. Schumann, J.; Hertweck, C. Molecular basis of cytochalasan biosynthesis in fungi: Gene cluster analysis and evidence for the involvement of a PKS-NRPS hybrid synthase by RNA silencing. *J. Am. Chem. Soc.* **2007**, *129*, 9564–9565. [CrossRef] [PubMed]

10. Oh, H.; Swenson, D.C.; Gloer, J.B.; Wicklow, D.T.; Dowd, P.F. Chaetochalasin A: A new bioactive metabolite from *Chaetomium brasiliense*. *Tetrahedron. Lett.* **1998**, *39*, 7633–7636. [CrossRef]

11. Zhang, Y.; Wang, T.; Pei, Y.H.; Hua, H.M.; Feng, B.M. Aspergillin PZ, a novel isoindole-alkaloid from *Aspergillus awamori*. *J. Antibiot.* **2002**, *55*, 693–695. [CrossRef] [PubMed]

12. Lin, Z.J.; Zhu, T.J.; Wei, H.J.; Zhang, G.J.; Wang, H.; Gu, Q.Q. Spicochalasin A and New Aspochalasins from the Marine-Derived Fungus *Spicaria elegans*. *Eur. J. Org. Chem.* **2009**, *18*, 3045–3051. [CrossRef]

13. Zhu, H.C.; Chen, C.M.; Tong, Q.Y.; Li, X.N.; Yang, J.; Xue, Y.B.; Luo, Z.W.; Wang, J.P.; Yao, G.M.; Zhang, Y.H. Epicochalasines A and B: Two bioactive merocytochalasans bearing caged epicoccine dimer units from *Aspergillus flavipes*. *Angew. Chem. Int. Ed.* **2016**, *55*, 3486–3490. [CrossRef] [PubMed]

14. Li, X.; Zhao, Z.; Ding, W.; Ye, B.; Wang, P.; Xu, J. Aspochalazine A, a novel polycyclic aspochalasin from the fungus *Aspergillus* sp. Z4. *Tetrahedron Lett.* **2017**, *58*, 2405–2408. [CrossRef]

15. Liu, Y.; Zhao, S.; Ding, W.; Wang, P.; Yang, X.; Xu, J. Methylthio-aspochalasins from a marine-derived fungus *Aspergillus* sp. *Mar. Drugs* **2014**, *12*, 5124–5131. [CrossRef] [PubMed]

16. Xu, J.Z.; Zhao, S.Z.; Yang, X.W. A new cyclopeptide metabolite of marine gut fungus from *Ligia oceanica*. *Nat. Prod. Res.* **2014**, *28*, 994–997. [CrossRef] [PubMed]

17. Wang, P.M.; Zhao, S.Z.; Liu, Y.; Ding, W.J.; Qiu, F.; Xu, J.Z. Asperginine, an unprecedented alkaloid from the marine-derived fungus *Aspergillus* sp. *Nat. Prod. Commun.* **2015**, *10*, 1363–1364. [PubMed]

18. Wang, T.; Zhang, Y.; Pei, Y.h. Two novel trichothecenes from *Myrothecium roridum*. *Med. Chem. Res.* **2007**, *16*, 155–161. [CrossRef]

19. Ding, G.; Wang, H.; Li, L.; Chen, A.J.; Chen, L.; Chen, H.; Zhang, H.; Liu, X.; Zou, Z. Trichoderones A and B: Two pentacyclic cytochalasans from the plant endophytic fungus *Trichoderma gamsii*. *Eur. J. Org. Chem.* **2012**, *2012*, 2516–2519. [CrossRef]

20. Liu, R.; Gu, Q.; Zhu, W.; Cui, C.; Fan, G.; Fang, Y.; Zhu, T.; Liu, H. 10-Phenyl-[12]-cytochalasins Z7, Z8, and Z9 from the marine-derived fungus *Spicaria elegans*. *J. Nat. Prod.* **2006**, *69*, 871–875. [CrossRef] [PubMed]

21. Chen, C.; Tong, Q.; Zhu, H.; Tan, D.; Zhang, J.; Xue, Y.; Yao, G.; Luo, Z.; Wang, J.; Wang, Y.; et al. Nine new cytochalasan alkaloids from Chaetomium globosum TW1-1 (Ascomycota, Sordariales). *Sci. Rep.* **2016**, *6*, 18711. [CrossRef] [PubMed]

22. Zheng, C.J.; Shao, C.L.; Wu, L.Y.; Chen, M.; Wang, K.L.; Zhao, D.L.; Sun, X.P.; Chen, G.Y.; Wang, C.Y. Bioactive phenylalanine derivatives and cytochalasins from the soft coral-derived fungus, *Aspergillus elegans*. *Mar. Drugs* **2013**, *11*, 2054–2068. [CrossRef] [PubMed]

23. Tomikawa, T.; Shin-ya, K.; Kinoshita, T.; Miyajima, A.; Seto, H.; Hayakawa, Y. Selective cytotoxicity and stereochemistry of aspochalasin D. *J. Antibiot.* **2001**, *54*, 379–381. [CrossRef] [PubMed]

24. Rukachaisirikul, V.; Rungsaiwattana, N.; Klaiklay, S.; Phongpaichit, S.; Borwornwiriyapan, K.; Sakayaroj, J. γ-Butyrolactone, cytochalasin, cyclic carbonate, eutypinic acid, and phenalenone derivatives from the soil fungus *Aspergillus* sp. PSU-RSPG185. *J. Nat. Prod.* **2014**, *77*, 2375–2382. [CrossRef] [PubMed]

25. Lin, Z.J.; Zhu, T.J.; Chen, L.; Gu, Q.Q. Three new aspochalasin derivatives from the marine-derived fungus *Spicaria elegans*. *Chin. Chem. Lett.* **2010**, *21*, 824–826. [CrossRef]

26. Wang, F.; Hua, H.M.; Pei, Y.H.; Chen, D.; Jing, Y.K. Triterpenoids from the resin of *Styrax tonkinensis* and their antiproliferative and differentiation effects in human leukemia HL-60 cells. *J. Nat. Prod.* **2006**, *69*, 807–810. [CrossRef] [PubMed]

 *marine drugs*

Article

Meroterpenoids and Isocoumarinoids from a *Myrothecium* Fungus Associated with *Apocynum venetum*

Yanchao Xu [1,2,3,†], Cong Wang [2,4,†], Haishan Liu [2], Guoliang Zhu [2], Peng Fu [2,3], Liping Wang [1,2,*] and Weiming Zhu [1,2,3,*]

1   State Key Laboratory of Functions and Applications of Medicinal Plants, Guizhou Medical University, Guiyang 550014, China; m18586818694@163.com
2   Key Laboratory of Marine Drugs, Ministry of Education of China, School of Medicine and Pharmacy, Ocean University of China, Qingdao 266003, China; wangcong123206@163.com(C.W.); liuhaishan_229@outlook.com (H.L.); guoliangzhu2015@hotmail.com (G.Z.); fupeng@ouc.edu.cn (P.F.)
3   Laboratory for Marine Drugs and Bioproducts, Qingdao National Laboratory for Marine Science and Technology, Qingdao 266003, China
4   Guangxi Key Laboratory of Chemistry and Engineering of Forest Products, School of Chemistry and Chemical Engineering, Guangxi University for Nationalities, Nanning 530006, China
*   Correspondence: lipingw2006@163.com (L.W.); weimingzhu@ouc.edu.cn (W.Z.); Tel./Fax: +86-532-8203-1268 (W.Z.)
†   These authors contributed equally to this paper.

Received: 31 August 2018; Accepted: 27 September 2018; Published: 1 October 2018

Abstract: Four new meroterpenoids **1–4** and four new isocoumarinoids **5–8**, along with five known isocoumarinoids (**9–13**), were isolated from the fungus *Myrothecium* sp. OUCMDZ-2784 associated with the salt-resistant medicinal plant, *Apocynum venetum* (Apocynaceae). Their structures were elucidated by means of spectroscopic analysis, X-ray crystallography, ECD spectra and quantum chemical calculations. Compounds **1–5**, **7**, **9** and **10** showed weak α-glucosidase inhibition with the IC$_{50}$ values of 0.50, 0.66, 0.058, 0.20, 0.32, 0.036, 0.026 and 0.37 mM, respectively.

Keywords: endophytic fungus; *Myrothecium* sp.; meroterpenoids; isocoumarinoids; α-glucosidase inhibitors; salt-resistant plant; *Apocynum venetum*

1. Introduction

Since the discovery of penicillin, fungi have been an important source of lead compounds for drug development, which have provided a lot of attractive natural products (NPs) with different biological activities [1–3]. With the increase of study on the terrestrial fungal NPs, more and more known compounds were isolated repeatedly. Therefore, many researchers turned their attention to the fungi isolated from specific habitats, such as the marine-derived fungi [4–7] and the fungi associated with the plants or animals [8–11].

As part of our ongoing studies to search for bioactive NPs from fungi derived from special niche [12–16], we screened the fungus *Myrothecium* sp. OUCMDZ-2784 which is associated with the salt-resistant plant *A. venetum* (Apocynaceae) growing in the Yellow River Delta, a traditional Chinese medicine used for treatment of hypertension [17] and heart failure [18]. *Myrothecium* sp. has been reported to produce trichothecenes [19], sesquiterpenes [20,21], diterpenes [22] and cyclopeptides [23] with cytotoxic and antibacterial activities. The ethyl acetate (EtOAc) extract of the fermentation of *Myrothecium* sp. OUCMDZ-2784 showed 75% inhibition of α-glucosidase at 286 μg/mL. Chemical study resulted in the isolation and identification of four new meroterpenoids, myrothecisins A–D (**1–4**) and four new isocoumarinoids, myrothelactones A–D (**5–8**), together with five known isocoumarinoids that were identified as tubakialactone B (**9**) [24], acremonone G (**10**) [25], 6,8-dihydroxy-3-methylisocoumarin (**11**) [26], 3,4-dimethyl-6,8-dihydroxyisocoumarin (**12**) [27] and sescandelin B (**13**) [28], respectively by comparing ^{1}H and ^{13}C NMR spectra (Figure S57, Table S1) as well as ESIMS spectra (Figure S59) with those reported.

2. Results and Discussion

Myrothecisin A (**1**) was isolated as a pale-yellow oil. Its molecular formula was assigned as $C_{25}H_{34}O_7$ by the HRESIMS peak at *m/z* 469.2188 [M + Na]$^+$ (Figure S58A), indicating nine degrees of unsaturation. The ^{13}C NMR (Figure S6) spectrum of **1** showed 25 signals that were classified by DEPT (Figure S7) and HSQC (Figure S8) as an aldehyde carbonyl carbon (δ_C 193.8), one acyl carbonyl carbon (δ_C 170.3), five sp^2 non-protonated carbons (δ_C 167.3, 159.8, 149.5, 112.3, 111.2) and three sp^3 non-protonated carbons (δ_C 98.8, 42.8, 39.2), one sp^2 methine (δ_C 101.5) and four sp^3 methines (δ_C 78.2, 71.4, 45.6, 36.0), five sp^3 methylenes (δ_C 60.5, 35.1, 30.6, 30.4, 20.7) and five methyl carbons (δ_C 28.7, 21.3, 17.0, 16.6, 15.2) (Table 1). The ^{1}H (Figure S5) and HSQC NMR showed the singlet signals at δ_H 10.06 and 6.56 for an aldehyde proton and an aromatic proton, respectively. The key HMBC (Figure S10) correlations from H-7' (δ_H 10.06) to C-2'/C-3'/C-4', 2'-OH (δ_H 12.16) to C-1'/C-2'/C-3', H-5' (δ_H 6.56) to C-1'/C-4'/C-6'/C-8', H-8' (δ_H 4.74) to C-3'/C-4'/C-5' and 8'-OH (δ_H 5.44) to C-4'/C-8' suggested a penta-substituted benzene ring (Figure 2). The COSY (homonuclear correlation spectroscopy) correlations from H-1 through H-2 to H-3 and H-5 through H-6, H-7 and H-8 to H-12 (Figure 1 and Figure S9), along with the key HMBC correlations from H-2 to C-4/C-10/C-16, H-3 to C-5/C-13/C-14, 3-OH to C-2/C-3/C-4, H-1 to C-2/C-3/C-10/C-15, H-15 to C-1/C-5/C-9/C-10, H-5 to C-3/C-4/C-6/C-13, H-6 to C-5, H-7 to C-8, H-13 to C-3/C-5/C-14, H-14 to C-3/C-5/C-13, H-12 to C-7/C-8/C-9 and H-17 to C-16 revealed a sesquiterpene fragment (Figure 2). The connection of the above-mentioned two fragments were confirmed by the key HMBC correlations from H-11 to C-8/C-9/C-10/C-1'/C-2'/C-6' (Figure 2) [29]. The relative configuration of **1** was determined by the NOESY correlations from H-8 to H-11 and H-15, H-3 to H-5 and H-13, H-2 to H-14 and H-15 and H-11 to H-15 (Figure 3 and Figure S11). The absolute configuration of **1** was determined by calculation of electronic circular dichroism (ECD) using time-dependent density functional theory (TDDFT) (Figure S1) [30,31] and the measured ECD spectrum of **1** matched well with the calculated ECD spectrum for (2*R*,3*R*,5*S*,8*R*,9*R*,10*S*)-**1** (Figure 4).

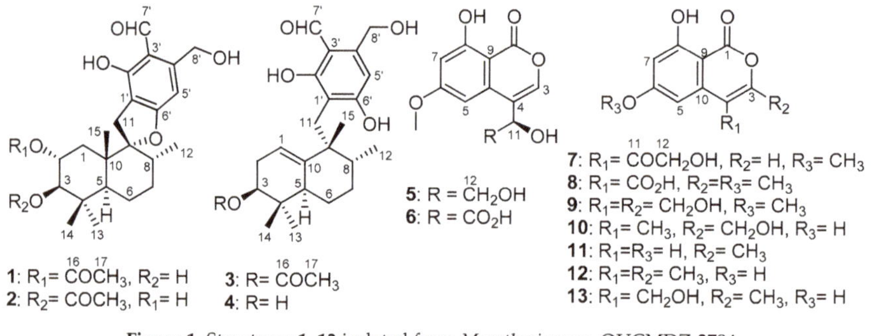

Figure 1. Structures **1–13** isolated from *Myrothecium* sp. OUCMDZ-2784.

The molecular formula of **2** was also determined as $C_{25}H_{34}O_7$ by the HRESIMS peak at *m/z* 469.2200 [M + Na]$^+$ (Figure S58B), implying that **2** is an isomer of **1**. Comparison of its ^{1}H and ^{13}C NMR spectra (Figures S12–S16) with those of **1** revealed that the acetyloxy group in **2** was on C-3. This was confirmed by the HMBC (Figure S17) correlation from H-3 (δ_H 4.29) to C-16 (δ_C 170.3) (Figure 2). The similar NOESY correlations (Figure 3 and Figure S18) suggested that **2** has the same relative configuration as **1**. The similarity of ECD curves between **2** and **1** (Figure 5) indicated the same absolute configurations of its stereogenic carbons. Therefore **2** was named myrothecisin B.

Table 1. ^{1}H (600 MHz) and ^{13}C (150 MHz) NMR data for **1–4** in DMSO-d_6.

No.	1		2		3		4	
	δ_C, Type	δ_H, Mult. (J in Hz)	δ_C, Type	δ_H, Mult. (J in Hz)	δ_C, Type	δ_H, Mult. (J in Hz)	δ_C, Type	δ_H, Mult. (J in Hz)
1	35.1, CH$_2$	1.16, m; 1.55 [a]	38.4, CH$_2$	1.17, m; 1.57 [a]	114.7, CH	4.94, brs	116.3, CH	4.88, s
2	71.4, CH	4.83, ddd (11.5, 10.1, 4.1)	64.4, CH	3.65, m	28.0, CH$_2$	1.89, m; 1.23, m	31.6, CH$_2$	1.72, m
3	78.2, CH	2.96, dd (10.1, 4.8)	83.3, CH	4.29, d (9.9)	75.6, CH	4.62, dd (8.6, 6.5)	71.5, CH	3.30 [a]
4	39.2, C		39.0, C		35.6, C		36.9, C	
5	45.6, CH	1.56 [a]	45.5, CH	1.63 [a]	43.1, CH	2.54, m	43.7, CH	2.45, m
6	20.7, CH$_2$	1.54 [a]; 1.47, m	20.4, CH$_2$	1.56 [a]; 1.48, m	26.7, CH$_2$	1.79, m	26.7, CH$_2$	1.77, m
7	30.6, CH$_2$	1.55 [a]; 1.35, m	30.5, CH$_2$	1.57 [a]; 1.38, m	30.2, CH$_2$	1.54, m	30.3, CH$_2$	1.52, m
8	36.0, CH	1.83, m	35.6, CH	1.85, m	43.3, CH	1.30, m	42.9, CH	1.29, m
9	98.8, C		98.8, C		43.7, C		43.7, C	
10	42.8, C		42.5, C		143.9, C		143.0, C	
11	30.4, CH$_2$	3.03, d (16.1); 2.79, d (16.0)	30.2, CH$_2$	3.11, d (16.5); 2.81, d (16.5)	23.5, CH$_2$	2.68, d (12.9); 2.53, d (12.9)	23.3, CH$_2$	2.63, d (12.7); 2.50, d (12.7)
12	15.2, CH$_3$	0.64, d (6.3)	15.1, CH$_3$	0.68, d (6.3)	17.1, CH$_3$	1.03, d (6.6)	17.2, CH$_3$	1.02, d (6.7)
13	28.7, CH$_3$	0.97, s	28.4, CH$_3$	0.79, s	25.0, CH$_3$	0.88, s	25.2, CH$_3$	0.95, s
14	17.0, CH$_3$	0.78, s	17.5, CH$_3$	0.81, s	17.1, CH$_3$	0.73, s	14.8, CH$_3$	0.59, s
15	16.6, CH$_3$	1.04, s	16.6, CH$_3$	1.03, s	23.0, CH$_3$	0.96, s	22.9, CH$_3$	0.91, s
16	170.3, C		170.3, C		169.8, C			
17	21.3, CH$_3$	1.91, s	21.0, CH$_3$	2.02, s	21.0, CH$_3$	1.98, s		
1'	111.2, C		111.2, C		111.4, C		111.3, C	
2'	159.8, C		159.8, C		164.5, C		164.8, C	
3'	112.3, C		112.2, C		110.2, C		110.2, C	
4'	149.5, C		149.4, C		145.4, C		145.4, C	
5'	101.5, CH	6.56, s	101.6, CH	6.61, s	107.5, CH	6.51, s	107.3, CH	6.50, s
6'	167.3, C		167.3, C		164.6, C		164.6, C	
7'	193.8, CH	10.06, s	193.8, CH	10.07, s	193.3, CH	9.96, s	193.4, CH	9.96, s
8'	60.5, CH$_2$	4.74, s	60.4, CH$_2$	4.74, s	59.9, CH$_2$	4.70, s	59.8, CH$_2$	4.70, s
2/3-OH		4.94, d (4.7)		4.62, d (4.9)				4.13, s
2'-OH		12.16, s		12.24, s		12.84, s		12.78, s
6'-OH								10.55, brs
8'-OH		5.44, s		5.39, s		5.35, s		5.36, s

[a] Overlapped.

89

Compound **3** was also obtained as a pale-yellow oil. Its molecular formula was determined as $C_{25}H_{34}O_6$ according to the HRESIMS peak at *m/z* 453.2238 [M + Na]$^+$ (Figure S58C). The ^{13}C NMR (Figure S20) spectrum of **3** showed one aldehyde carbonyl carbon (δ_C 193.3), one acyl carbonyl carbon (δ_C 169.8), six sp^2 non-protonated carbons (δ_C 164.6, 164.5, 145.4, 143.9, 111.4, 110.2) and two sp^3 non-protonated carbons (δ_C 43.7, 35.6), two sp^2 methines (δ_C 114.7, 107.5) and three sp^3 methines (δ_C 75.6, 43.3, 43.1), five sp^3 methylenes (δ_C 59.9, 30.2, 28.0, 26.7, 23.5) and five methyl carbons (δ_C 25.0, 23.0, 21.0, 17.1, 17.1) (Table 1). Analysis of its 1D and 2D NMR (Figures S19–S24) data revealed the presence of a substituted benzene ring and a sesquiterpene unit, indicating **3** was an analogue of **1** and **2**. Comparison of the ^{1}H and ^{13}C NMR spectra with those of **1** and **2** suggested a same pentasubstituted benzene ring. The structure of the sesquiterpene unit was slightly modified and was determined by the COSY (Figure S23) correlations from H-1 through H-2 to H-3 and H-5 through H-6, H-7 and H-8 to H-12 and the key HMBC (Figure S24) correlations from H-3 to C-2/C-4/C-10/C-13/C-14, H-17 to C-16, H-1 to C-3/C-9, H-15 to C-10, H-8 to C-9, H-5 to C-9/C-13, H-12 to C-7 and H-2 to C-4/C-10 (Figure 2). The HMBC correlations from H$_2$-11 (δ_H 2.68/2.53) to C-8/C-10/C-15/C-2′/C-6′ (Figure 2) confirmed the connection between the sesquiterpene fragment and the benzene ring. The relative configuration of **3** was determined by the NOESY (Figure S25) correlations from H-13 to H-3 and H-5, H-8 to H-15, as well as H-5 to H-11 (Figure 3). The absolute configuration was determined as (3*S*,5*R*,8*R*,9*R*)- by comparison of the calculated and experimental ECD spectra (Figure 4 and Figure S2). Therefore **3** was named myrothecisin C.

The molecular formula of **4** was assigned as $C_{23}H_{32}O_5$ by the HRESIMS peak at *m/z* 411.2139 [M + Na]$^+$ (Figure S58D), which was C_2H_2O less than that of **3**. The similarity of the UV and NMR data between **3** and **4** (Table 1) suggested that **4** possesses the same skeleton as **3**. Careful comparison of their ^{1}H and ^{13}C NMR spectra (Figures S26–S31) showed that the acetyloxy group (δ_C 21.0/δ_H 1.98 and δ_C 169.8) in **3** was replaced by a hydroxy group (δ_H 4.13) in **4** (Table 1). The NOESY data (Figure 3 and Figure S32) suggested that **4** has the same relative configuration as **3**. The ECD Cotton effects of **4** were nearly identical to those of **3** (Figure 5), indicating the same absolute configurations of the corresponding stereogenic carbons. Thus, **4** was named myrothecisin D.

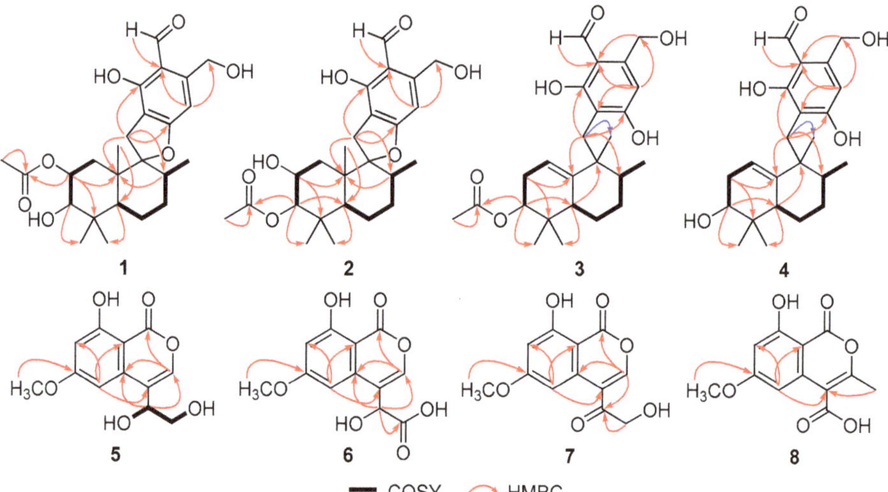

Figure 2. Key homonuclear correlation spectroscopy (COSY) and key HMBC correlations for **1–8**.

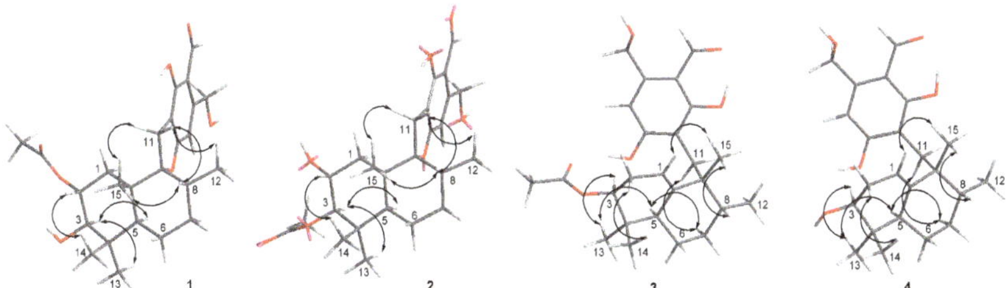

Figure 3. NOESY correlations for **1–4**.

Compound **5** was obtained as a colorless crystal with the molecular formula $C_{12}H_{12}O_6$ from the HRESIMS peak at *m/z* 251.0563 [M − H]⁻ (Figure S58E). The ¹H NMR spectrum showed two *meta*-coupled aromatic protons at δ_H 6.78 (d, *J* = 2.2 Hz) and δ_H 6.63 (d, *J* = 2.2 Hz) (Table 2, Figure S33), indicating the presence of a tetra-substituted benzene ring. The ¹³C (Figure S34) NMR spectrum showed 12 carbon signals that were classified by DEPT (Figure S35) and HSQC (Figure S36) spectra as six sp² non-protonated carbons (δ_C 166.4, 165.4, 163.4, 137.6, 118.6, 99.9), three sp² methines (δ_C 143.1, 100.5, 100.4) and one sp³ methine (δ_C 68.8), one sp³ methylene (δ_C 64.8) and one methoxy group (δ_C 56.0) (Table 2). The key HMBC correlations (Figure 2 and Figure S38) from CH₃O-6 to C-6, HO-11 to C-4/C-12, H-11 to C-3/C-10, H-3 to C-1/C-10, H-5 to C-4/C-7/C-9 and H-7 to C-9 along with the continuous COSY correlations of HO-11 (δ_H 5.50)/H-11 (δ_H 4.66)/H-12 (δ_H 3.51, 3.62)/HO-12 (δ_H 4.81) (Figure S37) revealed that **5** possesses a 4,6,8-trisubstituted isocoumarin skeleton with a hydroxy, a methoxy and a 1,2-dihydroxy ethyl at C-8, C-6 and C-4, respectively. The structure of **5** was further confirmed by X-ray crystallography (Figure 6). Because the value of the Flack parameter [−0.2(2)] was large, the absolute configuration determined by X-ray crystallography was not reliable. Thus, the ECD calculation method was used to further confirm the absolute configuration of C-11 of **5** as 11*R*- (Figure 4 and Figure S3). Consequently, **5** was named myrothelactone A.

Compound **6** was obtained as a white powder. Its molecular formula was determined as $C_{12}H_{10}O_7$ based on the HRESIMS peak at *m/z* 265.0355 [M − H]⁻ (Figure S58F). The UV and ¹³C NMR data of **6** (Table 2) were similar to those of **5**, indicating that they have the same isocoumarin scaffold. Comparison of their ¹H and ¹³C data (Figures S39–S43) indicated that the hydroxymethyl group ($\delta_{C/H}$ 64.8/3.62&3.51, δ_H 4.81) in **5** was replaced by the carboxyl group (δ_C 173.4). This change was verified by the key HMBC (Figure S44) correlations from H-11 to C-3/C-10/C-12. The absolute configuration of C-11 of **6** was determined as 11*R*- by comparison of the calculated and experimental ECD spectra (Figure 4 and Figure S4). Therefore, **6** was name myrothelactone B.

Table 2. ^{1}H and ^{13}C NMR data for **5–8** in DMSO-d_6.

No.	5 [a]		6 [b]		7 [b]		8 [b]	
	δ_C, Type	δ_H, mult. (J in Hz)	δ_C, Type	δ_H, mult. (J in Hz)	δ_C, Type	δ_H, mult. (J in Hz)	δ_C, Type	δ_H, mult. (J in Hz)
1	165.4, C		164.9, C		163.4, C		165.2, C	
3	143.1, CH	7.45, s	144.4, CH	7.62, s	153.1, CH	8.47, s	151.0, C	
4	118.6, C		117.7, C		114.7, C		118.7, C	
5	100.5, CH	6.78, d (2.2)	101.6, CH	6.84, s	102.3, CH	7.46, s	101.6, CH	6.86, s
6	166.4, C		166.4, C		166.8, C		166.3, C	
7	100.4, CH	6.63 (d, 2.2)	100.6, CH	6.63, s	101.2, CH	6.68, s	99.8, CH	6.48, s
8	163.4, C		163.3, C		163.2, C		162.6, C	
9	99.9, C		99.9, C		99.9, C		99.2, C	
10	137.6, C		136.8, C		134.8, C		138.6, C	
11	68.8, CH	4.66, td (5.2, 5.2)	68.5, CH	5.01, s	198.3, C		173.9, C	
12	64.8, CH$_2$	3.62, ddd (11.3, 5.3, 5.3); 3.51, ddd (11.3, 5.3, 5.3)	173.4, C		65.9, CH$_2$	4.57, d (4.7)	17.9, CH$_3$	2.29, s
6-OCH$_3$	56.0, CH$_3$	3.88, s	56.0, CH$_3$	3.84, s	56.0, CH$_3$	3.87, s	55.8, CH$_3$	3.80, s
8-OH		11.37, s		11.22, s		11.03, s		11.21, s
11-OH		5.50, d (4.7)						
12-OH		4.81, t (5.2)				5.28, t (5.0)		

[a] Data were measured at 600 MHz (^{1}H) and 150 MHz (^{13}C). [b] Data were measured at 500 MHz (^{1}H) and 125 MHz (^{13}C).

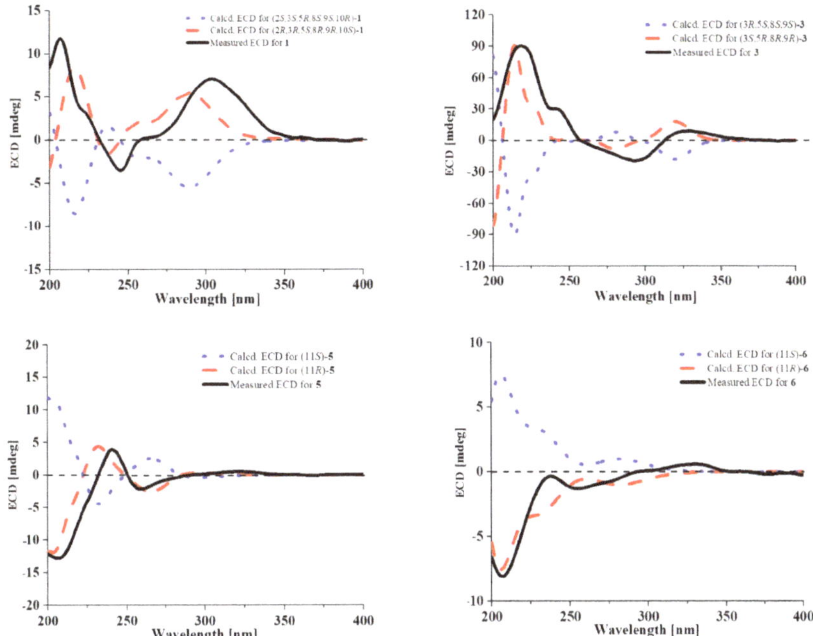

Figure 4. Measured and calculated ECD spectra for **1**, **3**, **5** and **6**.

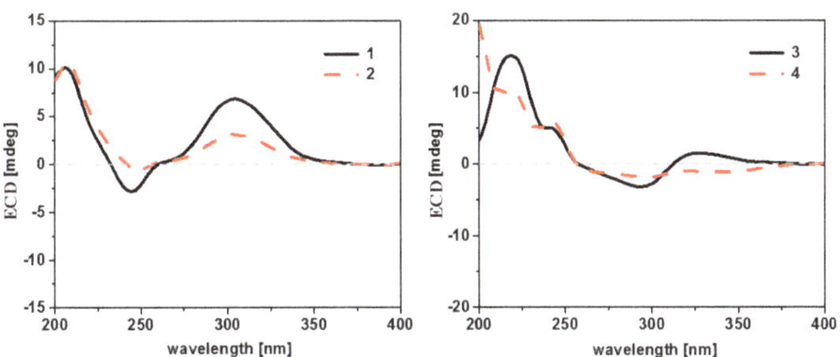

Figure 5. ECD spectra for **1−4**.

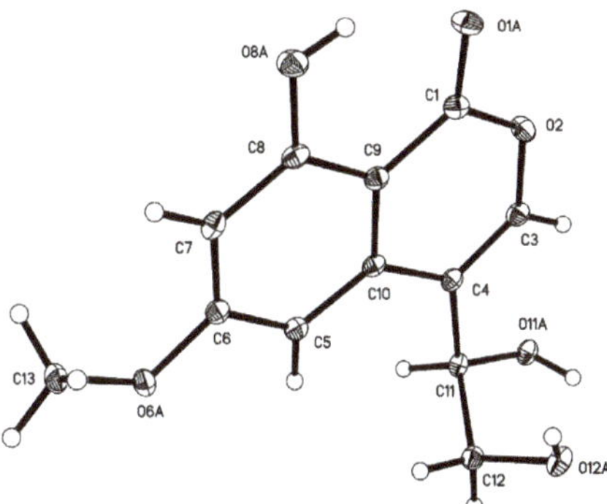

Figure 6. ORTEP diagram of **5**.

Compound **7** was obtained as a white powder. Its molecular formula was determined as $C_{12}H_{10}O_6$ according to its HRESIMS peak at *m/z* 249.0408 [M – H]$^-$ (Figure S58G), which was only two hydrogen atoms less than that of **5**. The difference observed in the NMR spectra of **7** and **5** was that the signals for hydroxymethine ($\delta_{C/H}$ 68.8/4.66) in **5** were replaced by the signal of a carbonyl group ($\delta_{C\text{-}11}$ 198.3) in **7** (Table 2, Figures S45–S49). The HMBC (Figure S50) correlations from H-3 (δ_H 8.47) and H-12 (δ_H 4.57) to C-11 further confirmed the structure of **7** which was name myrothelactone C (Figure 2).

The molecular formula of compound **8** was determined as $C_{12}H_{10}O_6$ on the basis of its HRESIMS peak at *m/z* 249.0407 [M − H]$^-$ (Figure S58H), which is an isomer of **7**. Analysis of its ^{1}H and ^{13}C NMR spectra showed that **8** also had the same isocoumarin scaffold, whose difference is the replacement of carbonyl (δ_C 198.3), hydroxymethyl (δ_C 65.9, δ_H 4.57/5.28) and sp^2 methine (δ_C 153.1, δ_H 8.47) signals in **7** by two sp^2 non-protonated carbons (δ_C 173.9, 151.0) and methyl (δ_C 17.9, δ_H 2.29) signals in **8** (Table 2, Figures S51–S55). The HMBC correlations from H-12 (δ_H 2.29) to C-3 (δ_C 151.0) and C-4 (δ_C 118.7) suggested the methyl substitution at C-3 (Figure 2 and Figure S56). The chemical shift of the carboxyl signal (δ_C 173.9) together with 2D NMR data indicated the carboxyl substitution at C-4. The structure of myrothelactone D (**8**) was therefore determined (Figure 2).

The α-glucosidase inhibitory activity of **1–13** was preliminarily investigated. Compounds **1–5**, **7**, **9** and **10** exhibited inhibitory activity against the human-sourced α-glucosidase recombinant expressed in *Saccharomyces cerevisiae* [31–33] with IC$_{50}$ values of 0.50, 0.66, 0.058, 0.20, 0.32, 0.036, 0.026 and 0.37 mM, while the IC$_{50}$ value of positive control acarbose was 0.47 mM. Due to the low activity, the deeper investigation of the mechanism and type of enzymatic inhibition as well as the binding mode were not done.

3. Experimental Section

3.1. General Experimental Procedures

Optical rotations were measured using a JASCO P-1020 digital polarimeter (JASCO Corporation, Tokyo, Japan). UV spectra were obtained on a Beckman DU 640 spectrophotometer (Beckman Coulter, Inc., Brea, CA, USA). CD data were performed using a JASCO J-815 spectropolarimeter (JASCO Corporation, Tokyo, Japan). IR spectra were obtained on a Nicolet Nexus 470 spectrophotometer (Thermo Nicolet Corporation, Madison, WI, USA) as KBr discs. NMR spectra were recorded on a Varian System 500 spectrometer (Varian, Palo Alto, CA, USA) or a Bruker Avance 600 spectrometer (Bruker, Fallanden, Switzerland) using residual solvent signals for referencing and chemical shifts were recorded as δ values. HRESIMS spectra were measured using the Q-TOF ULTIMA GLOBAL GAA076 LC mass spectrometer (Waters Asia, Ltd., Singapore). Semi-preparative HPLC was performed using an ODS column (YMC-pack ODS-A, 10 mm × 250 mm, 5 μm, 4.0 mL/min, Kyoto, Japan). TLC and column chromatography (CC) were performed on plates pre-coated with silica gel GF$_{254}$ (10–40 μm, Qingdao Marine Chemical Factory, Qingdao, China) and Sephadex LH-20 (Amersham Biosciences, Uppsala, Sweden), respectively. Vacuum-liquid chromatography (VLC) utilized silica gel H (Qingdao Marine Chemical Factory).

3.2. Collection and Phylogenetic Analysis

The fungus OUCMDZ-2784 was isolated from *Apocynum venetum* (Apocynaceae) collected from the estuary of Yellow River, Dongying, China. The leaves of the plant were washed with tap water and sterile distilled water in sequence. Then, it was cut into small pieces, which were then put into a centrifuge tubes filled with different concentrations of sucrose solution. These tubes were centrifuged at 1200 rpm for 20 min. Four zones were separated by improved discontinuous sucrose gradient centrifugation. The interface between the third and the fourth bands was deposited on a PDA (200 g potato, 20 g glucose, 20 g agar per liter of sea water) plate containing chloramphenicol (100 μg/mL) as a bacterial inhibitor and was then cultured at 28 °C for 3 days. A single colony was transferred to PDA agar media and was identified as *Myrothecium* sp. according to its morphological characteristics and 18S rRNA gene sequences (GenBank accession No. KF977010).

3.3. Cultivation and Extraction

Fungus OUCMDZ-2784 was prepared on PDA agar medium. Spores were incubated at 28 °C for 48 h on a rotary shaker with shaking at 120 rpm in a 500 mL cylindrical flask containing 150 mL liquid medium (20 g maltose, 20 g mannitol, 10 g glucose, 3 g yeast extract, 10 g monosodium glutamate per liter of sea water). The cultures were transferred to 350 × 1000 mL Erlenmeyer flasks and each containing 300 mL liquid fermentation media (1 g peptone, 10g soluble starch per liter of sea water, pH 7.0). The flasks were incubated at room temperature under static conditions for 30 days. The cultures were extracted three times by EtOAc and the combined EtOAc extracts were dried in vacuo to yield 20.1 g of extract.

3.4. Purification

The extract (20.1 g) was fractionated by VLC, eluting with a step gradient of CH_2Cl_2-petroleum ether (50–100%) and MeOH-CH_2Cl_2 (0–50%) and five fractions (Fr.1–Fr.5) were collected. Fraction 2 (3.2 g) was subjected to Sephadex LH-20 chromatography eluting with CH_2Cl_2/MeOH (1:1) to afford three subfractions (Fr.2.1–Fr.2.3). Fr.2.1 (1.0 g) was further purified by HPLC on an ODS column (80% MeOH/H_2O) to give compounds **1** (25.2 mg, t_R 6.3 min) and **3** (30.1 mg, t_R 10.2 min). Fr.2.2 (50.2 mg) was purified by HPLC on an ODS column (60% MeOH/H_2O) to yield compounds **11** (3.5 mg, t_R 10.4 min) and **8** (3.3 mg, t_R 15.2 min). Fr.2.3 (46.3 mg) was purified by HPLC on an ODS column (60% MeOH/H_2O) to yield compounds **12** (5.1 mg, t_R 12.6 min) and **13** (12.0 mg, t_R 16.4 min). Fraction 3 (3.8 g) was separated into three subfractions (Fr.3.1–Fr.3.3) by Sephadex LH-20 eluting with MeOH-CH_2Cl_2 (1:1). Fr.3.1 (0.5 g) was purified by semi preparative HPLC on an ODS column (85% MeOH/H_2O) to yield compound **2** (36.2 mg, t_R 10.2 min). Fr.3.2 (1.1 g) was separated by silica gel VLC column eluting with CH_2Cl_2-petroleum (2:1) to yield compounds **10** (100.3 mg) and **5** (200.8 mg). Fr.3.3 (0.5 g) was further purified by Sephadex LH-20 eluting with MeOH to yield compound **9** (15.3 mg). Fraction 4 (1.8 g) was separated into three subfractions (Fr.4.1–Fr.4.3) by Sephadex LH-20 eluting with MeOH. Fr.4.1 (0.6 g) was further purified by semi preparative HPLC on an ODS column (85% MeOH/H_2O) to yield compound **4** (34.3 mg, t_R 12.2 min). Fr.4.2 (0.2 g) was further purified by semi preparative HPLC on an ODS column (50% MeOH/H_2O) to yield compound **7** (6.0 mg, t_R 12.7 min). Fr.4.3 (0.3 g) was purified by HPLC on an ODS column (40% MeOH/H_2O) to yield compound **6** (5.8 mg, t_R 10.6 min).

Myrothecisin A (**1**): pale yellow oil; $[\alpha]_D^{20}$ + 9.0 (*c* 0.1, CHCl$_3$); UV (MeOH) λ_{max} (log ε) 212 (4.13), 237 (3.60), 305 (3.93) nm; ECD (0.002 M, MeOH) λ_{max} ($\Delta\varepsilon$) 207 (+1.69), 244 (−0.51), 304 (+1.06) nm; IR (KBr) ν_{max} 3443, 2926, 1719, 1635, 1372, 1267 cm^{-1}; ^{1}H and ^{13}C NMR data, see Table 1; HRESIMS *m/z* 469.2188 [M + Na]$^+$ (calcd for $C_{25}H_{34}O_7Na$, 469.2197).

Myrothecisin B (**2**): pale yellow oil; $[\alpha]_D^{20}$ + 70.1 (*c* 0.1, CHCl$_3$); UV (MeOH) λ_{max} (log ε) 211 (4.10), 238 (3.62), 305 (3.88) nm; ECD (0.002 M, MeOH) λ_{max} ($\Delta\varepsilon$) 207 (+1.62), 249 (−0.09), 305 (+0.49) nm; IR (KBr) ν_{max} 3424, 2940, 1733, 1638, 1460, 1371, 1263 cm^{-1}; ^{1}H and ^{13}C NMR data, see Table 1; HRESIMS *m/z* 469.2200 [M + Na]$^+$ (calcd for $C_{25}H_{34}O_7Na$, 469.2197).

Myrothecisin C (**3**): pale yellow oil; $[\alpha]_D^{20}$ + 63.5 (*c* 0.1, CHCl$_3$); UV (MeOH) λ_{max} (log ε) 212 (4.09), 239 (3.65), 307 (3.81) nm; ECD (0.002 M, MeOH) λ_{max} ($\Delta\varepsilon$) 219 (+2.29), 293 (−0.49), 327 (+0.22) nm; IR (KBr) ν_{max} 3480, 2936, 1732, 1612, 1373, 1254 cm^{-1}; ^{1}H and ^{13}C NMR data, see Table 1; HRESIMS *m/z* 453.2238 [M + Na]$^+$ (calcd for $C_{25}H_{34}O_6Na$, 453.2248).

Myrothecisin D (**4**): pale yellow oil; $[\alpha]_D^{20}$ + 20.0 (*c* 0.1, CHCl$_3$); UV (MeOH) λ_{max} (log ε) 203 (4.16), 224 (3.56), 300 (3.86) nm; ECD (0.002 M, MeOH) λ_{max} ($\Delta\varepsilon$) 220 (+1.57), 242 (+0.92), 295 (−0.27) nm; IR (KBr) ν_{max} 2928, 1718, 1621, 1370, 1264, 1027 cm^{-1}; ^{1}H and ^{13}C NMR data, see Table 1; HRESIMS *m/z* 411.2139 [M + Na]$^+$ (calcd for $C_{23}H_{32}O_5Na$, 411.2142).

Myrothelactone A (**5**): colorless crystal; melting point (mp) 174–175 °C; $[\alpha]_D^{20}$ − 36.0 (*c* 0.1, MeOH); UV (MeOH) λ_{max} (log ε) 246 (3.82), 328 (3.19) nm; ECD (0.004 *M*, MeOH) λ_{max} ($\Delta\varepsilon$) 206.5 (−0.22), 241

(+0.24), 259 (−0.27) nm; IR (KBr) ν_{max} 3747, 3630, 3159, 2959, 1668, 1558, 1398, 1237 cm^{-1}; ^{1}H and ^{13}C NMR data, see Table 2; HRESIMS *m/z* 251.0563 [M − H]$^-$ (calcd for $C_{12}H_{11}O_6$, 251.0561).

Myrothelactone B (**6**): white powder; mp 169–171 °C; $[\alpha]_D^{20}$ − 30.0 (*c* 0.1, MeOH); UV (MeOH) λ_{max} (log ε) 245 (3.81), 328 (3.17) nm; ECD (0.004 M, MeOH) λ_{max} ($\Delta\varepsilon$) 206.5 (−0.78), 241 (+0.24), 259 (−0.13) nm; IR (KBr) ν_{max} 3749, 2922, 1681, 1651, 1619, 1459, 1399 cm^{-1}; ^{1}H and ^{13}C NMR data, see Table 2; HRESIMS *m/z* 265.0355 [M − H]$^-$ (calcd for $C_{12}H_9O_7$, 265.0354).

Myrothelactone C (**7**): white powder; mp 160–161 °C; UV (MeOH) λ_{max} (log ε) 228 (3.91), 263 (3.43), 325 (3.16) nm; IR (KBr) ν_{max} 3750, 3675, 3615, 1736, 1651, 1558, 1540, 1399 cm^{-1}; ^{1}H and ^{13}C NMR data, see Table 2; HRESIMS *m/z* 249.0408 [M − H]$^-$ (calcd for $C_{12}H_9O_6$, 249.0405).

Myrothelactone D (**8**): white powder; mp 219–221 °C; UV (MeOH) λ_{max} (log ε) 231 (3.92), 263 (3.45), 325 (3.16) nm; IR (KBr) ν_{max} 3749, 3673, 3445, 3197, 1716, 1682, 1539, 1457, 1399 cm^{-1}; ^{1}H and ^{13}C NMR data, see Table 2; HRESIMS *m/z* 249.0407 [M − H]$^-$ (calcd for $C_{12}H_9O_6$, 249.0405).

3.5. X-ray Structure Determination of Compound 5

Compound **5** was obtained as a colorless needles crystal with molecular formula $C_{12}H_{12}O_6$. Orthorhombic, space group $P2_12_12_1$, a = 4.9041(2) Å, b = 13.8470(5) Å, c = 15.7443(6) Å, α = 90.00°, β = 90.00°, γ = 90.00°, V = 1069.15(7) Å3, Z = 4, D_{calcd} = 1.567 Mg/m^3, μ = 1.089 mm^{-1}, $F(000)$ = 528, crystal size 0.30 mm × 0.18 mm × 0.15 mm, T = 293(2) K. A total of 1478 unique reflections ($2\theta < 50°$) were collected on a CCD area detector diffractometer with graphite monochromated Cu Kα radiation (λ = 1.54178 Å). The structure was solved by direct methods (SHELXS-97) and expanded using Fourier techniques (SHELXL-97). The final cycle of full-matrix least squares refinement was based on 1478 unique reflections ($2\theta < 50°$) and 165 variable parameters and converged with unweighted and weighted agreement factors of R_1 = 0.0326, wR_2 = 0.0885 and R = 0.0880 for $I > 2\text{sigma}(I)$ data. Absolute structure parameter: −0.2(2). The deposited number of compound **5** in the Cambridge Crystallographic Data Centre is 980155.

3.6. α-Glucosidase Inhibitory Assays

The human-sourced α-glucosidase was recombinant expressed in the yeast *Saccharomyces cerevisiae* and the inhibitory effects of compounds **1–13** were tested using p-nitrophenyl-α-D-glucopyranoside (pNPG) as substrate [31–33]. The sample was dissolved in sodium phosphate buffer (PBS, pH 6.8) at three concentrations. 10 μL of the sample solution, 20 μL of 2.5 mM pNPG solution (in phosphate buffer) and 20 μL of PBS were mixed in a 96-well microplate and incubated at 37 °C for 5 min. A volume of 10 μL of α-glucosidase diluted to 0.2 U/mL by 0.01 M PBS was then added to each well. After incubating at 37 °C for 15 min, the absorbance at 405 nm was recorded by a Spectra max 190 micro plate reader (Molecular Devices Inc., San Jose, CA, USA). The blank was prepared by adding phosphate buffer instead of the α-glucosidase and the positive control was acarbose. Blank readings (no enzyme) were subtracted from each well and results were compared to the control. The inhibition (%) was calculated as $[1 − (OD_{drug}/OD_{blank})] \times 100\%$. The IC$_{50}$ value was calculated as the compound concentration that is required for 50% inhibition and the IC$_{50}$ value of acarbose was 0.47 mM.

4. Conclusions

This study revealed eight new fungal NPs, meroterpenoids **1–4** and isocoumarinoids **5–8**, from the culture of the salt-tolerant plant-associated fungus *Myrothecium* sp. OUCMDZ-2784. The new compounds **1–5** and **7** exhibited α-glucosidase inhibitory activity. Combined with bioactive NPs from mangrove-derived fungi [34–38], the results indicated that fungi living in the salt-tolerant plants are an important biological resources for new and bioactive NPs.

Supplementary Materials: The following are available online at http://www.mdpi.com/1660-3397/16/10/363/s1, Figures S1–S4: DFT-optimized structures for low-energy conformers of compounds **1**, **3**, **5** and **6**, Figures S5–S11: NMR spectra of compound **1** in DMSO-d_6, Figures S12–S18: NMR spectra of compound **2** in DMSO-d_6, Figures S19–S25: NMR spectra of compound **3** in DMSO-d_6, Figures S26–S32: NMR spectra of compound **4** in DMSO-d_6, Figures S33–S38: NMR spectra of compound **5** in DMSO-d_6, Figures S39–S44: NMR spectra of compound **6** in DMSO-d_6, Figures S45–S50: NMR spectra of compound **7** in DMSO-d_6, Figures S51–S56: NMR spectra of compound **8** in DMSO-d_6, Figure S57: ^{1}H- and ^{13}C-NMR spectra of compounds **9–13** in DMSO-d_6, Figure S58: HRESI-MS spectra of compounds **1–8**, Figure S59: ESI-MS spectra of compounds **9–13**, Table S1: ^{1}H and ^{13}C NMR data for compounds **9–13** in DMSO-d_6.

Author Contributions: Y.X. performed the most experiments; C.W. prepared the draft of the manuscript; H.L. tested the α-glucosidase inhibitory activity; G.Z. performed the ECD calculations; P.F. checked the data; L.W. revised the manuscript; W.Z. designed and supervised the research and revised the final version.

Funding: This research was funded by the NSFC (Nos. 81561148012, U1501221, 81741150, U1606403), the 100 Leading Talents of Guizhou Province (fund for W. Zhu), the science and technology project of Guizhou (Grant No. QKHT Z-2014-4007) and the academician workstation of Guizhou (Grant No. QKH YSZ-2015-4009).

Conflicts of Interest: The authors declare no conflict of interest.

References

1. Aly, A.H.; Debbab, A.; Proksch, P. Fifty years of drug discovery from fungi. *Fungal Divers.* **2011**, *50*, 3–19. [CrossRef]

2. Bladt, T.T.; Frisvad, J.C.; Knudsen, P.B.; Larsen, T.O. Anticancer and antifungal compounds from *Aspergillus*, *Penicillium* and other filamentous fungi. *Molecules* **2013**, *18*, 11338–11376. [CrossRef] [PubMed]

3. Khan, A.A.; Bacha, N.; Ahmad, B.; Lutfullah, G.; Farooq, U.; Cox, R.J. Fungi as chemical industries and genetic engineering for the production of biologically active secondary metabolites. *Asian Pac. J. Trop. Biomed.* **2014**, *4*, 859–870. [CrossRef]

4. Rateb, M.E.; Ebel, R. Secondary metabolites of fungi from marine habitats. *Nat. Prod. Rep.* **2011**, *28*, 290–344. [CrossRef] [PubMed]

5. Bugnia, T.S.; Ireland, C.M. Marine-derived fungi: A chemically and biologically diverse group of microorganisms. *Nat. Prod. Rep.* **2004**, *21*, 143–163. [CrossRef] [PubMed]

6. Xu, L.; Meng, W.; Cao, C.; Wang, J.; Shan, W.; Wang, Q. Antibacterial and antifungal compounds from marine fungi. *Mar. Drugs* **2015**, *13*, 3479–3513. [CrossRef] [PubMed]

7. Moghadamtousi, S.Z.; Nikzad, S.; Kadir, H.A.; Abubakar, S.; Zandi, K. Potential antiviral agents from marine fungi: An overview. *Mar. Drugs* **2015**, *13*, 4520–4538. [CrossRef] [PubMed]

8. Strobel, G.; Daisy, B.; Castillo, U.; Harper, J. Natural products from endophytic microorganisms. *J. Nat. Prod.* **2004**, *67*, 257–268. [CrossRef] [PubMed]

9. Borges, W.S.; Borges, K.B.; Bonato, P.S.; Said, S.; Pupo, M.T. Endophytic fungi: natural products, enzymes and biotransformation reactions. *Curr. Org. Chem.* **2009**, *13*, 1137–1163. [CrossRef]

10. Nisa, H.; Kamili, A.N.; Nawchoo, I.A.; Shafi, S.; Shameem, N.; Bandh, S.A. Fungal endophytes as prolific source of phytochemicals and other bioactive natural products: A review. *Microb. Pathog.* **2015**, *82*, 50–59. [CrossRef] [PubMed]

11. Tan, R.X.; Zou, W.X. Endophytes: A rich source of functional metabolites. *Nat. Prod. Rep.* **2001**, *18*, 448–459. [CrossRef] [PubMed]

12. Sun, K.; Li, Y.; Guo, L.; Wang, Y.; Liu, P.; Zhu, W. Indole diterpenoids and isocoumarin from the fungus, *Aspergillus flavus*, isolated from the prawn, *Penaeus vannamei*. *Mar. Drugs* **2014**, *12*, 3970–3981. [CrossRef] [PubMed]

13. Zhu, G.; Kong, F.; Wang, Y.; Fu, P.; Zhu, W. Cladodionen, a cytotoxic hybrid polyketide from the marine-derived *Cladosporium* sp. OUCMDZ-1635. *Mar. Drugs* **2018**, *16*, 71. [CrossRef] [PubMed]

14. Wang, Y.; Zheng, J.; Liu, P.; Wang, W.; Zhu, W. Three new compounds from *Aspergillus terreus* PT06-2 grown in a high salt medium. *Mar. Drugs* **2011**, *9*, 1368–1378. [CrossRef] [PubMed]

15. Wang, Y.; Lu, Z.; Sun, K.; Zhu, W. Effects of high salt stress on secondary metabolites from marine-derived fungus *Spicaria elegans*. *Mar. Drugs* **2011**, *9*, 535–542. [CrossRef] [PubMed]

16. Wang, Y.; Wang, L.; Zhuang, Y.; Kong, F.; Zhang, C.; Zhu, W. Phenolic polyketides from the co-cultivation of marine-derived *Penicillium* sp. WC-29-5 and *Streptomyces fradiae* 007. *Mar. Drugs* **2014**, *12*, 2079–2088. [CrossRef] [PubMed]

17. Chinese Pharmacopeia Committee of Ministry of Public Health of the People's Republic of China. *Chinese Pharmacopeia 2000*; Chemical and Technical Press: Beijing, China, 2000; p. 170.

18. Shaanxi Provincial and Municipal Collaborative Group for Prevention and Treatment of Coronary Heart Diseases. Clinical observation and pharmacological experiment of *Apocynum venetum* root in the treatment of heart failure. *Shaanxi Med. J.* **1974**, *5*, 10–14.

19. Isaka, M.; Punya, J.; Lertwerawat, Y.; Tanticharoen, M.; Thebtaranonth, Y. Antimalarial activity of macrocyclic trichothecenes isolated from the fungus *Myrothecium verrucaria*. *J. Nat. Prod.* **1999**, *62*, 329–331. [CrossRef] [PubMed]

20. Fu, Y.; Wu, P.; Xue, J.; Wei, X. Cytotoxic and antibacterial quinone sesquiterpenes from a *Myrothecium* fungus. *J. Nat. Prod.* **2014**, *77*, 1791–1799. [CrossRef] [PubMed]

21. Fu, Y.; Wu, P.; Xue, J.; Li, H.; Wei, X. Myrothecols G and H, two new analogues of the marine-derived quinone sesquiterpene Penicilliumin A. *Mar. Drugs* **2015**, *13*, 3360–3367. [CrossRef] [PubMed]

22. Hsu, Y.H.; Hirota, A.; Shima, S.; Nakagawa, M.; Nozaki, H.; Tada, T.; Nakayama, M. Structure of myrocin C, a new diterpene antibiotic produced by a strain of *Myrothecium* sp. *Agric. Biol. Chem.* **1987**, *51*, 3455–3457. [CrossRef]

23. Zou, X.; Niu, S.; Ren, J.; Li, E.; Liu, X.; Che, Y. Verrucamides A–D, antibacterial cyclopeptides from *Myrothecium verrucaria*. *J. Nat. Prod.* **2011**, *74*, 1111–1116. [CrossRef] [PubMed]

24. Nakashima, K.; Tomida, J.; Hirai, T.; Morita, Y.; Kawamura, Y.; Inoue, M. Tubakialactones A–E, new polyketides from the endophytic fungus *Tubakia* sp. ECN-111. *Tetrahedron Lett.* **2017**, *58*, 2248–2251. [CrossRef]

25. Rukachaisirikul, V.; Rodglin, A.; Sukpondma, Y.; Phongpaichit, S.; Buatong, J.; Sakayaroj, J. Phthalide and isocoumarin derivatives produced by an *Acremonium* sp. isolated from a mangrove *Rhizophora apiculata*. *J. Nat. Prod.* **2012**, *75*, 853–858. [CrossRef] [PubMed]

26. Kornsakulkarn, J.; Thongpanchang, C.; Lapanun, S.; Srichomthong, K. Isocoumarin glucosides from the scale insect fungus *Torrubiella tenuis* BCC 12732. *J. Nat. Prod.* **2009**, *72*, 1341–1343. [CrossRef] [PubMed]

27. Tanaka, A.K.; Sato, C.; Shibata, Y.; Kobayashi, A.; Yamashita, K. Growth promoting activities of sclerotinin A and its analogs. *Agric. Biol. Chem.* **1974**, *38*, 1311–1315. [CrossRef]

28. Kimura, Y.; Nakadoi, M.; Nakajima, H.; Hamasaki, T.; Nagai, T.; Kohmoto, K.; Shimada, A. Structure of sescandelin-B, a new metabolite produced by the fungus *Sesquicillium candelabrum*. *Agric. Biol. Chem.* **1991**, *55*, 1887–1888.

29. Zhao, J.; Feng, J.; Tan, Z.; Liu, J.; Zhao, J.; Chen, R.; Xie, K.; Zhang, D.; Li, Y.; Yu, L.; et al. Stachybotrysins A−G, phenylspirodrimane derivatives from the fungus *Stachybotrys chartarum*. *J. Nat. Prod.* **2017**, *80*, 1819–1826. [CrossRef] [PubMed]

30. Berova, N.; Bari, L.D.; Pescitelli, G. Application of electronic circular dichroism in configurational and conformational analysis of organic compounds. *Chem. Soc. Rev.* **2007**, *36*, 914–931. [CrossRef] [PubMed]

31. Kong, F.; Zhao, C.; Hao, J.; Wang, C.; Wang, W.; Huang, X.; Zhu, W. New α-glucosidase inhibitors from a marine sponge-derived fungus, *Aspergillus* sp. OUCMDZ-1583. *RSC Adv.* **2015**, *5*, 68852–68863. [CrossRef]

32. Wang, C.; Guo, L.; Hao, J.; Wang, L.; Zhu, W. α-Glucosidase inhibitors from the marine-derived fungus *Aspergillus flavipes* HN4-13. *J. Nat. Prod.* **2016**, *79*, 2977–2981. [CrossRef] [PubMed]

33. Shim, Y.-J.; Doo, H.-K.; Ahn, S.-Y.; Kim, Y.-S.; Seong, J.-K.; Park, I.-S.; Min, B.-H. Inhibitory effect of aqueous extract from the gall of *Rhus chinensis* on alpha-glucosidase activity and postprandial blood glucose. *J. Ethnopharmacol.* **2003**, *85*, 283–287. [CrossRef]

34. Lin, Z.; Zhu, T.; Fang, Y.; Gu, Q.; Zhu, W. Polyketides from *Penicillium* sp. JP-1, an endophytic fungus associated with the mangrove plant *Aegiceras corniculatum*. *Phytochemistry* **2008**, *69*, 1273–1278. [CrossRef] [PubMed]

35. Lu, Z.; Zhu, H.; Fu, P.; Wang, Y.; Zhang, Z.; Lin, H.; Liu, P.; Zhuang, Y.; Hong, K.; Zhu, W. Cytotoxic polyphenols from the marine-derived fungus *Penicillium expansum*. *J. Nat. Prod.* **2010**, *73*, 911–914. [CrossRef] [PubMed]

36. Wang, J.; Lu, Z.; Liu, P.; Wang, Y.; Li, J.; Hong, K.; Zhu, W. Cytotoxic polyphenols from the fungus *Penicillium expansum* 091006 endogenous with the mangrove plant *Excoecaria agallocha*. *Planta Med.* **2012**, *78*, 1861–1866. [PubMed]

37. Kong, F.; Wang, Y.; Liu, P.; Dong, T.; Zhu, W. Thiodiketopiperazines from the marine-derived fungus *Phoma* sp. OUCMDZ-1847. *J. Nat. Prod.* **2014**, *77*, 132–137. [CrossRef] [PubMed]

38. Wang, L.; Han, X.; Zhu, G.; Wang, Y.; Chairoungdua, A.; Piyachaturawat, P.; Zhu, W. Polyketides from the endophytic fungus *Cladosporium* sp. isolated from the mangrove plant *Excoecaria agallocha*. *Front. Chem.* **2018**, *6*, 344. [CrossRef] [PubMed]

Article

Anthraquinone Derivatives from a Marine-Derived Fungus *Sporendonema casei* HDN16-802

Xueping Ge [1], Chunxiao Sun [1], Yanyan Feng [1], Lingzhi Wang [1], Jixing Peng [3], Qian Che [1], Qianqun Gu [1], Tianjiao Zhu [1,2], Dehai Li [1,2] and Guojian Zhang [1,2,*]

[1] Key Laboratory of Marine Drugs, Chinese Ministry of Education, School of Medicine and Pharmacy, Ocean University of China, Qingdao 266003, China; 15610568273@163.com (X.G.); sunchunxiao93@163.com (C.S.); yy15715321143@163.com (Y.F.); wlz8866wlz@163.com (L.W.); cheqian064@ouc.edu.cn (Q.C.); guqianq@ouc.edu.cn (Q.G.); zhutj@ouc.edu.cn (T.Z.); dehaili@ouc.edu.cn (D.L.)

[2] Laboratory for Marine Drugs and Bioproducts of Qingdao National Laboratory for Marine Science and Technology, Qingdao 266237, China

[3] Key Laboratory of Testing and Evaluation for Aquatic Product Safety and Quality, Ministry of Agriculture and Rural Affairs, Yellow Sea Fisheries Research Institute, Chinese Academy of Fishery Sciences, Qingdao 266071, China; pengjixing1987@163.com

* Correspondence: zhangguojian@ouc.edu.cn; Tel.: +86-532-82032065

Received: 30 April 2019; Accepted: 29 May 2019; Published: 4 June 2019

Abstract: Five new anthraquinone derivatives, auxarthrols D–H (**1–5**), along with two known analogues (**6–7**), were obtained from the culture of the marine-derived fungus *Sporendonema casei*. Their structures, including absolute configurations, were established on the basis of NMR, HRESIMS, and circular dichroism (CD) spectroscopic techniques. Among them, compound **4** represents the second isolated anthraquinone derivative with a chlorine atom, which, with compound **6**, are the first reported anthraquinone derivatives with anticoagulant activity. Compounds **1** and **3** showed cytotoxic activities with IC_{50} values from 4.5 μM to 22.9 μM, while compounds **1**, **3–4**, and **6–7** showed promising antibacterial activities with MIC values from 12.5 μM to 200 μM. In addition, compound **7** was discovered to display potential antitubercular activity for the first time.

Keywords: anthraquinone derivatives; *Sporendonema casei*; marine-derived fungus; cytotoxic activities; antibacterial activities

1. Introduction

Anthraquinones and their derivatives are a group of pigmented polyketides widely produced by fungi. Apart from their bright color attributed to the typical conjugate system in their structure, they have also attracted the attention of scientists due to their diversity of structures and wide range of pharmacological effects, such as their anti-infective, anti-inflammatory, and α-glucosidase inhibitory activities and cytotoxicity against cancer cells [1,2]. Following the discovery of altersolanol A reported in 1967 [3], a series of anthraquinone derivatives have been discovered from various fungal genera, including *Alternaria* [4,5], *Streptomyces* [6,7], *Dactylaria* [8], *Bostryconema* [9], *Stemphylium* [10], *Pleospora* [11], *Auxarthron* [12], *Ampelomyces* [13], *Nigrospora* [14], and *Phomopsis* [15].

During our exploration of novel bioactive secondary metabolites obtained from marine-derived microorganisms, a fungus *Sporendonema casei* HDN16-802 isolated from a sediment sample collected from Zhangzi Island was selected due to its special morphological characteristic (orange color) and tremendous metabolic profile identified via HPLC-UV. Further chemical study generated five new anthraquinones, named auxarthrols D–H (**1–5**), along with two known analogues (**6–7**). To the best of our knowledge, this is the first time that anthraquinone derivatives have been isolated from the fungus *S. casei*. The cytotoxicity, antibacterial, anticoagulant, and antitubercular activities of **1–7** were tested. Herein, we will describe the isolation, structural elucidation, and biological activities of the isolated compounds.

2. Results and Discussion

Sporendonema casei HDN16-802 was cultured (45 L) under static conditions with oatmeal medium at room temperature for one month. The fermentation product (mycelium and broth) was extracted with ethyl acetate to provide the crude extract (10 g). The crude extract was fractionated by different kinds of chromatography, including silica gel vacuum liquid chromatography (VLC), C-18 column chromatography (ODS), Sephadex LH-20 column chromatography, medium performance liquid chromatography (MPLC), and finally HPLC to yield **1** (10.0 mg), **2** (2.1 mg), **3** (5.1 mg), **4** (5.0 mg), **5** (4.5 mg), **6** (10.1 mg), and **7** (5.0 mg) (Figure 1).

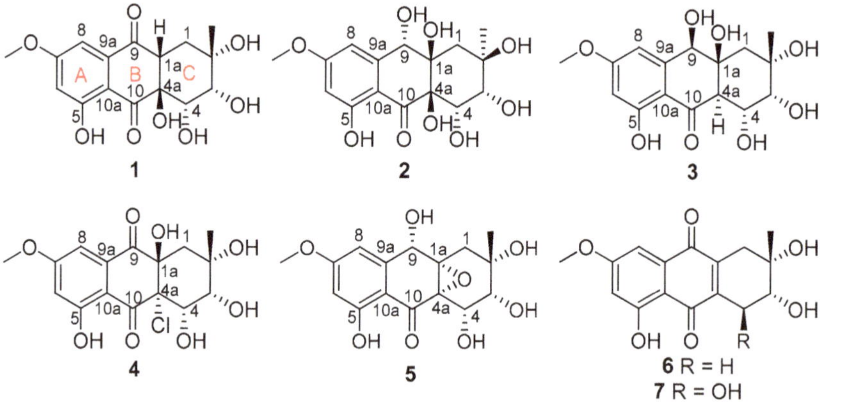

Figure 1. Structures of **1–7**.

Compound **1** was isolated as a pale yellow solid with the molecular formula $C_{16}H_{18}O_8$, which was established on the basis of the (+)–HRESIMS ion peak at *m/z* 339.1076 [M+H]$^+$ and *m/z* 361.0897 [M+Na]$^+$, indicating eight degrees of unsaturation. The 1D NMR (^{1}H-NMR, ^{13}C NMR, and DEPT) spectrum (Tables 1 and 2 and Supplementary Materials), together with HSQC correlations (Figure S5), provided five hydroxyl protons, including a chelated hydroxyl at δ_H 12.44 (s); two methyls, including one methoxy (δ_H 3.87, s; δ_C 56.7); one methylene (δ_H 1.86, m; δ_C 34.2); five methines, including two *meta*-coupled aromatic sp^2 methines [δ_H 6.79, d (2.4), δ_C 106.1; δ_H 6.83, d (2.4), δ_C 105.0] and two sp^3 oxygenated methines [δ_H 3.61, dd (5.7, 3.1), δ_C 73.7; δ_H 4.22, dd (4.5, 3.1), δ_C 71.3]; and eight non-protonated carbons, including two conjugated ketones (δ_C 197.3 and 200.1), four aromatic carbons (δ_C 166.1, 166.3, 110.1, and 137.6), and two oxygenated quaternary (δ_C 72.3 and 78.2) carbons. A careful comparison of the above signals with those of the known compound auxarthrol B [12] revealed a very similar hydroanthraquinone skeleton, while the most significant differences were the absence of a hydroxyl group and the appearance of a methine signal [δ_H 3.36, m; δ_C 48.0] attributed to C-1a (Table 2). The key HMBC correlations from H-1a to C-9 and C-2 (Figure 2 and Supplementary Materials) further confirmed the planar structure of **1**.

Table 1. ^{1}H NMR data of compounds **1–5** (500 MHz, TMS, δ ppm, *J* in Hz).

No.	1 [a]	2 [a]	2 [b]	3 [a]	4 [a]	5 [a]
1	1.86, m	1.92, d (14.6); 1.83, d (14.6)	1.92, d (14.6); 1.83, d (14.6)	1.82, d (14.2); 1.75, d (14.2)	2.24, d (14.8); 1.74, d (14.8)	2.35, dd (15.5); 2.31, dd (15.5)
3	3.61, dd (5.7, 3.1)	3.57, d (3.0)	3.57, m	3.46, t (3.0)	3.61, d (3.6)	3.46, d (3.7)
4	4.22, dd (4.5, 3.1)	4.43, d (3.0)	4.43, t (3.5)	4.39, dd (3.0, 9.7)	4.69, d (3.6)	4.57, d (3.7)
6	6.79, d (2.4)	6.35, d (2.5)	6.35, d (2.5)	6.34, d (2.4)	6.82, d (2.5)	6.42, d (2.4)
8	6.83, d (2.4)	6.64, d (2.5)	6.64, m	6.67, d (2.4)	6.96, d (2.5)	6.67, dd (2.4, 1.2)
9		4.73, s	4.73, d (9.5)	4.52, d (8.7)		4.83, d (1.2)
1a	3.36, (6.0, 1.5)					
4a				2.87, d (9.7)		
11	3.87, s	3.80, s	3.80, s	3.81, s	3.88, s	3.82, s
12	1.14, s	1.20, s	1.20, s	1.19, s	1.27, s	1.21, s
OH-2	4.28, s	5.69, s	5.69, s	5.44, s	6.24, s	
OH-3	4.43, d (5.7)		5.50, d (5.3)	4.81, d (3.0)	4.91, s	
OH-4	5.01, d (4.5)		4.52, d (3.5)	4.57, d (3.0)	4.96, s	
OH-4a	6.46, m	5.56, s	5.57, s			
OH-5	12.44, s	11.97, s	11.97, s	12.37, s	11.15, s	12.19, s
OH-1a		5.29, s	5.29, s	4.91, s	6.97, s	
OH-9			5.30, d (9.5)	5.46, d (8.7)		

[a] in DMSO; [b] in CDCl$_3$.

Table 2. ^{13}C NMR data of compounds **1–7** (125 MHz, DMSO, TMS, δ ppm).

No.	1	2	3	4	5
1	34.2, CH$_2$	34.3, CH$_2$	38.7, CH$_2$	30.7, CH$_2$	42.4, CH$_2$
2	72.3, C	73.5, C	73.2, C	73.6, C	70.3, C
3	73.7, CH	76.9, CH	75.2, CH	75.7, CH	72.9, CH
4	71.3, CH	64.6, CH	66.1, CH	63.5, CH	65.2, CH
5	166.1, C	164.8, C	163.8, C	163.6, C	164.9, C
6	106.1, CH	99.7, CH	99.5, CH	107.2, CH	100.1, CH
7	166.3, C	166.4, C	166.3, C	165.7, C	166.8, C
8	105.0, CH	106.2, CH	105.8, CH	106.4, CH	106.6, CH
9	197.3, C	70.0, CH	73.9, CH	190.6, C	68.6, CH
10	200.1, C	202.4, C	206.0, C	194.4, C	196.6, C
1a	48.0, CH	79.3, C	78.5, C	80.5, C	64.8, C
4a	78.2, C	78.1, C	53.3, CH	72.1, C	63.3, C
9a	137.6, C	148.4, C	149.0, C	134.5, C	145.9, C
10a	110.1, C	108.6, C	110.8, C	110.0, C	107.1, C
11	56.7, CH$_3$	56.1, CH$_3$	56.1, CH$_3$	56.7, CH$_3$	56.2, CH$_3$
12	23.1, CH$_3$	27.6, CH$_3$	27.8, CH$_3$	27.6, CH$_3$	26.1, CH$_3$

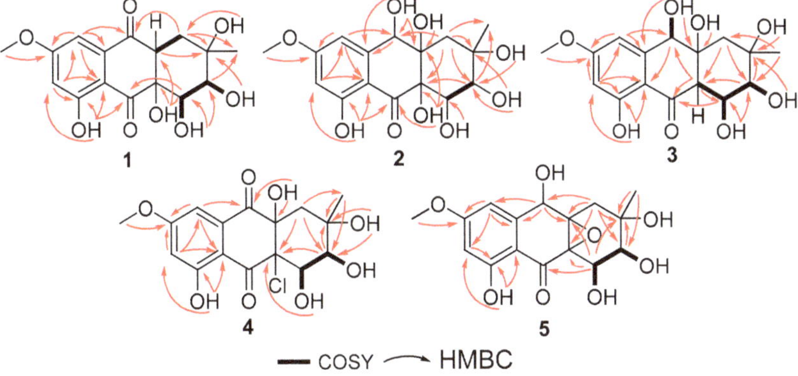

Figure 2. Key HMBC and ^{1}H-^{1}H COSY correlations of **1–5**.

The relative configuration of the stereogenic carbons in **1** was detected by NOESY correlations and conformational analysis (Figure 3 and Figure S7). The NOESY correlations from H-3 to H-1a and 4a-OH, from H-1a to H-4, and from H-4 to 4a-OH indicated that H-1a and 4a-OH were located on the same face of the molecule, which meant that the B ring and C ring were *cis* fused. The NOESY correlations from H-1a and H-4 to H_3-12 oriented H_3-12 to the same side as H-1a and H-4. Computational simulation by Chemdraw (Minimize Energy program), together with the small *J* coupling constant between H-3 and H-4 ($^3J_{H\text{-}3,\,H\text{-}4}$ = 3.1 Hz), further confirmed the chair-chair conformation for rings B and C, where H-1a (ax), H-3 (eq), H-4 (ax), 4a-OH (eq), and H_3-12 (ax) were oriented on the same face, thus completing the relative configuration of the stereogenic carbons in **1** (Figure 3). To determine the absolute configuration of compound **1**, the theoretical calculated electronic circular dichroism (ECD) spectra of possible models were performed using TDDFT. The optimized conformation of the model was obtained and further used for the ECD calculation at the B3LYP/6-31+G(d) level. The pattern (2*S*, 3*R*, 4*S*, 1a*R*, 4a*S*)-**1** of the calculated ECD spectrum was in reasonable agreement with the experimental ECD spectra (Figure 4). Thus, the absolute configuration of **1** was established as 2*S*, 3*R*, 4*S*, 1a*R*, 4a*S*, and we named it auxarthrol D.

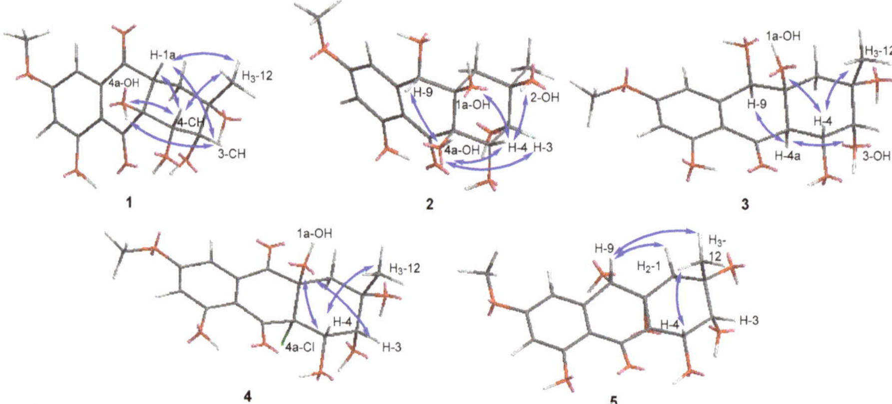

Figure 3. Key NOE correlations of **1–5**.

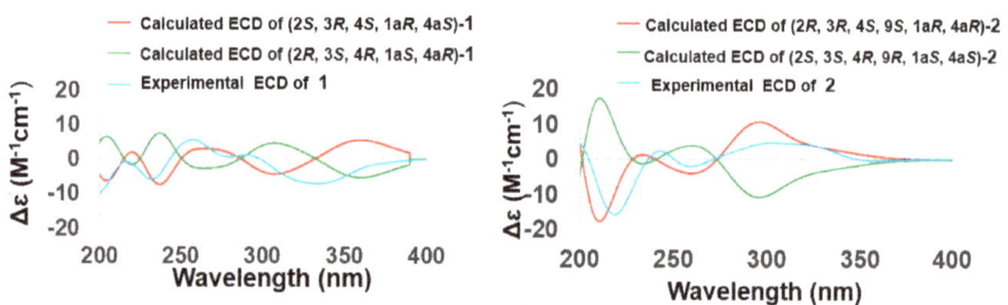

Figure 4. Comparison of the calculated and experimental ECD spectra of **1–2**, auxarthrols D–E.

Compound **2** was obtained as a pale yellow powder. Its molecular formula of $C_{16}H_{20}O_9$ with seven degrees of unsaturation was determined by the ion peak *m/z* 341.1237 $[M+H]^+$ in the (+)−HRESIMS. The molecular formula was also corroborated by exploiting ^1H and ^{13}C NMR spectroscopic data (Tables 1 and 2 and Supplementary Materials). A comparison of these data with **1** revealed the same skeleton with different substitutions. The upfield shift of C-9 (δ_C 69.0 Vs. δ_C 197.3) and the downfield shift of C-1a (δ_C 79.3 Vs. δ_C 48.0) indicated that both of the C-9 and C-1a positions were substituted by a hydroxyl group in **2** (Table 2). Key HMBC correlations from H-9 to C-9, C-8, and C-1; from 1a-OH to

C-1a and C-1; and from 9-OH to C-9 and C-9a (Figure 2 and Supplementary Materials) confirmed the locations of C-9 and C-1a hydroxyl groups, thus completing the planar structure of **2**. The relative configurations of the stereogenic carbons were determined by NOESY correlations and *J* coupling constant analysis (Table 1, Figure 3 and Supplementary Materials). The NOESY correlations from H-4 to both 1a-OH and 4a-OH and from H-3 to 4a-OH, together with the small *J* coupling constant between H-3 and H-4 ($^3J_{H-3, H-4}$ = 3.0 Hz), indicated that 1a-OH, 4a-OH, H-3, and H-4 were located on the same face. Other NOE correlations from H-3 to H_3-12 and 2-OH and from H-4 to 2-OH with the above evidence suggested a chair conformation of the C ring, where H-3 (eq), H-4 (ax), and 2-OH (ax) were on the same face, while H_3-12 (eq) was oriented on the opposite face of the molecule. This conformation was further confirmed by using Chemdraw Minimize Energy simulation. Further NOESY correlation from H-9 to 4a-OH indicated that H-9 was on the same face as 4a-OH (Figure 3), thus providing the relative configuration of the stereogenic carbons of **2**. The absolute configuration of **2** was determined by comparing the experimental and calculated ECD spectrum using time-dependent density-functional theory (TDDFT). The good agreement of the calculated ECD spectrum of (2*R*, 3*R*, 4*S*, 9*S*, 1a*R*, 4a*R*)-**2** with that of the experimental spectrum (Figure 4) suggested that the absolute configuration of **2** was 2*R*, 3*R*, 4*S*, 9*S*, 1a*R*, 4a*R*, and we named it auxarthrol E.

Compound **3** was obtained as a pale yellow powder. The molecular formula of **3** was deduced as $C_{16}H_{20}O_9$ (seven degrees of unsaturation) by (+)–HRESIMS *m/z* 357.1187 [M+H]$^+$, which was also corroborated by ^{1}H and ^{13}C NMR spectroscopic data, as shown in Tables 1 and 2, which was 16 amu more than the molecular mass of compound **2**, therefore revealing a close relationship between **3** and **2**. According to 1D NMR spectra, the presence of a methine signal at 2.87 ppm and the absence of a hydroxy group in **3** along with the upfield shift of C-4a (δ_C 53.3 Vs. δ_C 78.1) suggested that the 4a-OH in **2** was replaced by a hydrogen atom in **3** (Tables 1 and 2), which was confirmed by the key HMBC correlation from H-4a to C-10 and C-4 (Figure 2 and Supplementary Materials). The relative configurations of the stereogenic carbons were also determined by NOESY correlations and *J* coupling constant analysis. The NOESY correlations from H-9 to H-4a and the large *J* coupling constant between H-4 and H-4a ($^3J_{H-4, H-4a}$ = 9.7 Hz) indicated that H-4 (ax) and 1a-OH (ax) were on the same face, while H-4a (ax) was located on the opposite face, indicating that the B ring and C ring were *trans* fused. By using the Minimize Energy simulation programe in Chemdraw, both B and C rings were proposed to adopt a chair conformation, which provided the lowest steric energy. The NOESY correlation from H-4 to H_3-12 with the small *J* coupling constant between H-3 and H-4 ($^3J_{H-3, H-4}$ = 3.0 Hz) assigned H_3-12 and H-3 (eq) on the same face as H-4 (ax) (Figure 3), thus providing the relative configuration of the stereogenic carbons of **3**. The good agreement of the calculated ECD spectrum of (2*S*, 3*R*, 4*R*, 9*R*, 1a*S*, 4a*R*)-**3** with that of the experimental spectrum (Figure 5) suggested that the absolute configuration of **3** was 2*S*, 3*R*, 4*R*, 9*R*, 1a*S*, 4a*R*, and we named it auxarthrol F.

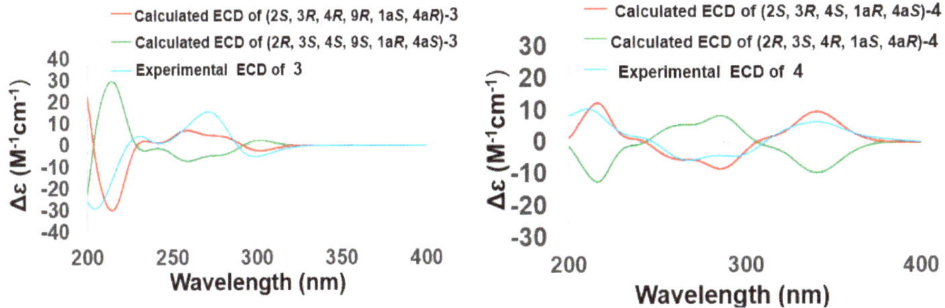

Figure 5. Comparison of the calculated and experimental ECD spectra of **3–4**, auxarthrols F–G.

Compound **4** was obtained as a pale yellow powder. Its molecular formula of $C_{16}H_{18}ClO_8$ (eight degrees of unsaturation) was determined by (+)–HRESIMS. The molecular formula was also corroborated by ^{1}H and ^{13}C NMR spectroscopic data (Tables 1 and 2), suggesting that the structure of **4** resembled that of paradictyoarthrin A (**8**) [16], except for the absence of the hydroxyl group on C-9 and the presence of a keto-carbonyl signal at δ_C 190.6, indicating that the C-9 hydroxyl was replaced by a ketone. Further 2D NMR data confirmed the planar structure of **4** (Figure 2 and Supplementary Materials). The relative configurations of the stereogenic carbons of **4** were established by NOESY correlations and *J* coupling constant analysis (Table 1, Figure 3 and Supplementary Materials). The NOESY correlations from H-4 and H-3 to 1a-OH indicated that 1a-OH and 4a-Cl were located on the opposite side of the B ring, while the NOESY correlation from H-4 to H_3-12 with a very small *J* coupling constant between H-3 and H-4 suggested that H-3 (eq), H-4 (ax), H_3-12 (ax), and 1a-OH (ax) were oriented on the same side of the C ring. Moreover, the calculated ECD spectrum of the model compound (2*S*, 3*R*, 4*S*, 1a*R*, 4a*S*)-**4** was well-matched with the experimental ECD spectrum of **4** (Figure 5), thus confirming the absolute structure of **4**, and we named it auxarthrol G.

Compound **5** was obtained as a pale yellow solid, with the molecular formula $C_{16}H_{18}O_8$ (eight degrees of unsaturation) from (+)–HRESIMS *m/z* 357.1187 [M+H]$^+$ combined with ^{1}H and ^{13}C NMR spectroscopic data (Tables 1 and 2). A comparison of 1D NMR data with those reported for altersolanol O [17] revealed a similar hydroanthraquinone skeleton, while the only differences were the absence of C-1 hydroxy and the replacement of the C-9 carbonyl group with a C-9 hydroxyl group in **5**, which was further confirmed by the key HMBC correlations from H-9 to C-1a and C-9a, and from H_2-1 to C-1a and C-9 (Figure 2 and Supplementary Materials). The relative configuration was also determined by NOESY correlations and *J* coupling constant analysis. The NOESY correlations from H-9 to H_3-12 and H_2-1 indicated that H-9 (ax) and H_3-12 (ax) were on the same face, showing that the B ring and C ring were *cis* fused with the C-1a and C-4a epoxide ring on the opposite side to H-9 (Figure 3). Further NOESY correlations from H-4 to 2-CH$_3$ with the small *J* coupling constant between H-3 and H-4 ($^3J_{\text{H-3, H-4}}$ = 3.7 Hz) suggested that H-3 (eq), H-4 (ax), and H_3-12 (ax) were on the same face of the C ring. The absolute configurations of the stereogenic carbons of **5** were determined as 2*S*, 3*R*, 4*S*, 1a*S*, 4a*S*, 9*S* by a comparison of the experimental and calculated ECD spectra (Figure 6). Compound **5** was named auxarthrol H.

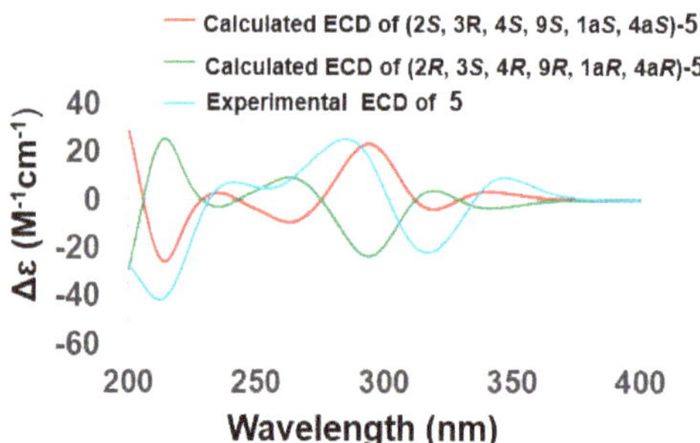

Figure 6. Comparison of the calculated and experimental ECD spectra of **5**, auxarthrol H.

By a comparison of the NMR and MS data with the literature, two known compounds were identified as 4-dehydroxyaltersolanol A (**6**) [14] and altersolanol B (**7**) [11] (Figure 1).

As a typical class of anthraquinone derivatives, auxarthrols characterized with multiple hydroxyl groups attached on the hydroanthraquinone skeleton were first isolated in 1969 [11] and this was followed by total synthesis and biosynthetic studies [18,19]. By the end of 2018, seventeen altersolanols [19] and three auxarthrols [12,19] had been discovered and because of their broad range of biological activities [20–22], this class of compounds has received growing attention from the natural product community.

Compounds **1**–**7** were tested for their cytotoxic activity against eleven types of human cancer cell lines using SRB staining [23] and MTT [24] methods, with doxorubicin hydrochloride (Dox) as a positive control. Compounds **1** and **3** showed moderate cytotoxic activity against eleven human cancer cell lines, with IC_{50} values ranging from 4.5 µM to 22.9 µM (Table 3). The antimicrobial activity of **1–7** was also evaluated and **1**, **3–4**, and **6–7** showed promising antibacterial activity, with MIC values ranging from 12.5 µM to 200 µM. (Table 4).

Table 3. Cytotoxic effect of **1** and **3** against eleven human cancer cell lines.

Comp.	IC$_{50}$ (µM)										
	HL-60	Hela	HCT-116	MGC-803	HO8910	MDA-MB-231	SH-SY5Y	PC-3	BEL-7402	K562	L-02
1	7.5	>50.0	14.5	21.8	>50.0	19.1	22.9	21.9	16.6	>50.0	>50.0
3	4.5	10.7	7.8	17.7	18.7	10.1	17.2	20.0	21.3	16.5	22.2
Dox [a]	0.1	0.6	0.2	0.2	0.4	0.2	0.1	1.0	0.4	0.3	0.4

[a] Dox stands for doxorubicin hydrochloride, which was used as a positive control.

Table 4. Antimicrobial effect of **1–7** on seven microorganisms.

Comp.	MIC (µM)						
	Mycobacterium Phlei	*Proteus Species*	*Bacillus subtilis*	*Candida albicans*	*Vibrio Parahemolyticus*	*Escherichia coli*	*Pseudomonas aeruginosa*
1	25.0	50.0	100	>200	50.0	100	50.0
2	>200	>200	>200	>200	>200	>200	>200
3	200	200	200	>200	>200	>200	200
4	50.0	25.0	25.0	200	100	>200	100
5	>200	>200	>200	>200	>200	>200	>200
6	25.0	50.0	25.0	>200	25.0	>200	25.0
7	25.0	100	25.0	>200	25.0	>200	12.5
Positive Control	3.12 [a]	1.56 [a]	0.781 [a]	1.56 [b]	0.781 [a]	0.391 [a]	1.56 [a]

[a] Ciprofloxacin used as a positive control for bacteria; [b] Nystatin used as a positive control for *Candida albicans*.

Moreover, all the compounds were investigated for their anticoagulant activity using argatroban as a positive control (inhibition ratio: 65.0%). Compounds **4** and **6** displayed a moderate effect with an inhibition ratio of 47.8% and 51.5%, respectively (Table 5). In addition, all the compounds were tested for antitubercular activity, but only **7** displayed a weak antitubercular effect, with an MIC value of 20.0 µg/mL (Table 6).

Table 5. Anticoagulant activity of **1–7**.

Comp.	1	2	3	4	5	6	7	Argatroban [b]
Inhibition ratio [a]	12.5	19.9	14.4	47.8	27.3	51.5	19.3	65.0

[a] Data are expressed as inhibition ratio values (%); [b] Argatroban was used as a positive control.

Table 6. Antitubercular activity of **1–7** against AlRa.

Comp.	1	2	3	4	5	6	7	Rifampin [b]
MIC [a]	>20.0	>20.0	>20.0	>20.0	>20.0	>20.0	20.0	1.0

[a] Data are expressed as MIC values (µg/mL); [b] Rifampin was used as a positive control.

3. Materials and Methods

3.1. General Experimental Procedures

UV spectra were recorded on Waters 2487. IR spectra were recorded on a Nicolet NEXUS 470 spectrophotometer in KBr discs (Thermo Scientific, Beijing, China). Optical rotations were measured on a JASCO P-1020 digital polarimeter (JASCO Corporation, Tokyo, Japan). HRESIMS and ESIMS data were obtained on a Thermo Scientific LTQ Orbitrap XL mass spectrometer. ECD spectra were measured on a JASCO J-715 spectra polarimeter (JASCO Corporation, Tokyo, Japan). NMR spectra were recorded on an Agilent 500 MHz DD2 spectrometer using TMS as the internal standard, and the chemical shifts were recorded as δ values. Semi-preparative HPLC was performed on an ODS column (HPLC (YMC-Pack ODS-A, 10 × 250 mm, 5 μm, 3 mL/min)). MPLC was performed on a Bona-Agela CHEETAHTM HP100 (Beijing Agela Technologies Co., Ltd., Beijing, China). Column chromatography (CC) was performed with silica gel (200–300 mesh, Qingdao Marine Chemical Inc. Qingdao, China) and Sephadex LH-20 (Amersham Biosciences, San Francisco, CA, USA), respectively.

3.2. Fungal Material

The fungal strain HDN16-802 was isolated from the sediment sample of Zhangzi Island, collected from Dalian, Liaoning Province, China. The strain was identified as *Sporendonema casei* based on sequencing of the ITS region (GenBank: MK578184). A voucher specimen strain was prepared on potato dextrose agar slants and deposited at $-20\ °C$ in the Key Laboratory of Marine Drugs, Chinese Ministry of Education.

3.3. Fermentation and Extraction

S. casei HDN16-802 was cultured on slants with PDA at 28 °C for 7 days. Further fermentation was carried out under static conditions at room temperature for 30 days in Erlenmeyer flasks (1000 mL), with each containing 53 g of oatmeal and naturally collected seawater (125 mL per flask) from Huiquan Bay, Qingdao, China. The pooled fermentation broth, together with mycelium (total of 45 L), were macerated and extracted with an equal volume of EtOAc three times. The organic layers were combined together and concentrated under reduced pressure to yield the extract (10 g).

3.4. Isolation

The extract (10 g) was fractionated by VLC column chromatography on silica gel using stepwise gradient elution with petroleum ether-CH_2Cl_2-MeOH (from PE only to PE with DCM in different ratios and DCM only later, and then from DCM only to DCM with MeOH in different ratios and MeOH only, depending on the polarity from small to large) to give six fractions (fraction 1 to fraction 6). Fraction 5 (eluted with 92:8 DCM-MeOH) was further separated by MPLC and then HPLC, eluting with MeOH/H_2O (35:65) to obtain **1** (t_R 28 min; 10.0 mg). Fraction 2 (eluted with 98:2 DCM-MeOH) was applied on a Sephadex LH-20 column and eluted with MeOH to provide six fractions (fraction 2-1 to fraction 2-6). Fraction 2-4 was separated by HPLC eluting with MeCN/H_2O (23:77) to obtain **4** (t_R 40 min; 5.0 mg) and **7** (t_R 35 min; 5.0 mg). Fraction 3 (eluted with 94:6 DCM-MeOH) was further separated by a C-18 ODS column with a step gradient elution of MeOH-H_2O (15:85-80:20), resulting in four fractions (fraction 3-1 to fraction 3-4). Fraction 3-1 was separated by HPLC eluting with MeCN/H_2O (gradient 15:85-25:75) to provide **2** (t_R 23 min; 2.1 mg), **3** (t_R 25 min; 5.1 mg), **5** (t_R 45 min; 4.5 mg), and **6** (t_R 50 min; 10.1 mg) and Fraction 3-1-3 was further purified by HPLC using MeOH/H_2O (24:76) as an eluent to obtain **5** (t_R 25 min; 7.9 mg).

Auxarthrol D (1): Pale yellow crystal; $[\alpha]_D^{25}$ −20.45 (c 0.03, MeOH); UV (MeOH) λ_{max} (log ε) 248 (2.4), 295 (1.2), 350 (1.2) nm; CD (2.5 mM, MeOH) λ_{max} (Δε) 218 (+0.74), 240 (−1.78) nm, 260 (+3.13)nm, 333 (−3.68) nm; IR (KBr) ν_{max} 3366, 2940, 2360, 1700, 1636, 1615, 1385, 1305, 1204, 1164, 1102, 1032, 910 cm^{-1}; for ^1H and ^{13}C NMR data, see Tables 1 and 2; HRESIMS m/z 339.1076 [M+H]$^+$ (calculated for $C_{16}H_{19}O_8$, 339.1074)

Auxarthrol E (2): Pale yellow powder; $[\alpha]_D^{25}$ −19.0 (c 0.04, MeOH); UV (MeOH) λ_{max} (log ε) 290 (2.4), 330 (1.6) nm; CD (2.5 mM, MeOH) λ_{max} (Δε) 218 (−1.74), 240 (+0.58) nm, 300 (+3.78) nm; IR (KBr) ν_{max} 3356, 2934, 2361, 1717, 1625, 1577, 1376, 1291, 1204, 1158, 1074, 851 cm^{-1}; for ^1H and ^{13}C NMR data, see Tables 1 and 2; HRESIMS m/z 357.1187 [M+H]$^+$ (calculated for $C_{16}H_{21}O_9$, 357.1180).

Auxarthrol F (3): Pale yellow powder; $[\alpha]_D^{25}$ −65.0 (c 0.2, MeOH); UV (MeOH) λ_{max} (log ε) 280 (2.4), 340 (1.0) nm; CD (2.5 mM, MeOH) λ_{max} (Δε) 218 (−1.54) nm, 240 (+0.78) nm, 300 (+3.12) nm; IR (KBr) ν_{max} 3379, 2925, 2362, 1626, 1375, 1296, 1204, 1160, 1084, 1032, 849, 780 cm^{-1}; for ^1H and ^{13}C NMR data, see Tables 1 and 2; HRESIMS m/z 341.1237 [M+H]$^+$ (calculated for $C_{16}H_{21}O_8$, 347.1231).

Auxarthrol G (4): Pale yellow powder; $[\alpha]_D^{25}$ +5.38 (c 0.08, MeOH); UV (MeOH) λ_{max} (log ε) 245 (2.4), 300 (1.0), 350 (1.2) nm; CD (2.5 mM, MeOH) λ_{max} ($\Delta\varepsilon$) 210 (+3.74), 267 (−1.78), 333 (2.73) nm; IR(KBr) ν_{max} 3375, 2924, 2359, 1636, 1615, 1385, 1296, 1205, 1162, 1030, 771 cm^{-1}; for ^{1}H and ^{13}C NMR data, see Tables 1 and 2; HRESIMS m/z 373.0685 [M+H]$^+$ (calculated for $C_{16}H_{16}O_8Cl$, 373.0685).

Auxarthrol H (5): Pale yellow powder; $[\alpha]_D^{25}$ +12.17 (c 0.2, MeOH); UV (MeOH) λ_{max} (log ε)274 (2.4), 315 (1.4) nm; CD (2.5 mM, MeOH) λ_{max} ($\Delta\varepsilon$) 218 (−1.74), 242 (+0.78) nm, 290 (+2.56) nm, 315 (−1.24) nm, 356 (+0.54) nm; IR (KBr) ν_{max} 3356, 2933, 2361, 1627, 1376, 1298, 1205, 1159, 1103, 1024, 950, 601 cm^{-1}; for ^{1}H and ^{13}C NMR data, see Tables 1 and 2; HRESIMS m/z 339.1080 [M+H]$^+$ (calculated for $C_{16}H_{19}O_8$, 339.1074).

3.5. Assay of Cytotoxic Activity

Cytotoxic activity was evaluated as previously reported [25].

3.6. Assay of Antimicrobial Activity

Antimicrobial activity was evaluated as previously reported [26].

3.7. Assay of Anticoagulant Activity

Anticoagulant activity was evaluated as previously reported [27].

3.8. Assay of Antitubercular Activity

Antitubercular activity was evaluated as previously reported [28].

4. Conclusions

In conclusion, we reported the isolation and structural elucidation of five new bioactive anthraquinone derivatives (**1–5**), together with two known analogues (**6–7**), from *Sporendonema casei*. Compound **4** is the second anthraquinone derivative with a chlorine atom. Compounds **1–7** were evaluated for their cytotoxic, antimicrobial, anticoagulant, and antitubercular activity. Compounds **1** and **3** showed cytotoxic activities against eleven human cancer cell lines, with IC$_{50}$ values ranging from 4.5 μM to 22.9 μM, while **1**, **3–4**, and **6–7** showed promising antibacterial activity, with MIC values ranging from 12.5 μM to 200 μM. Compounds **4** and **6** displayed a moderate anticoagulant effect, which are the first anthraquinone derivatives with this activity. In addition, **7** was found to display potential antitubercular activity.

Supplementary Materials: The following are available online at http://www.mdpi.com/1660-3397/17/6/334/s1, 1D and 2D NMR and HRESIMS spectra of **1–7**. Figures S1–S8: 1D and 2D NMR and HRESIMS spectra of Auxarthrol D (**1**); Figures S9–S17: 1D and 2D NMR and HRESIMS spectra of Auxarthrol E (**2**); Figures S18–S25: 1D and 2D NMR and HRESIMS spectra of Auxarthrol F (**3**); Figures S26–S33: 1D and 2D NMR and HRESIMS spectra of Auxarthrol G (**4**); Figures S34–S41: 1D and 2D NMR and HRESIMS spectra of Auxarthrol H (**5**).

Author Contributions: The contributions of the respective authors are as follows: X.G. drafted the work and performed the fermentation and extraction, as well as the isolation. X.G. and C.S. elucidated the constituents. Y.F. and L.W. were involved in the biological evaluations. J.P., Q.C., Q.G., T.Z., and D.L. contributed to checking and confirming all of the procedures of the isolation and identification. G.Z. designed the study, supervised the laboratory work, and contributed to the critical reading of the manuscript, and was also involved in structural determination and bioactivity elucidation. All the authors have read the final manuscript and approved the submission.

Funding: This work was financially supported by the National Natural Science Foundation of China (41806167, 41606166), the Marine S&T Fund of Shandong Province for the Pilot National Laboratory for Marine Science and Technology (Qingdao) (No. 2018SDKJ0401-2), the Fundamental Research Funds for the Central Universities (201941001), and the Taishan Scholar Youth Expert Program in Shandong Province (tsqn201812021).

Conflicts of Interest: The authors declare no conflict of interest.

References

1. Zhou, Y.X.; Xia, W.; Yue, W.; Peng, C.; Rahman, K.; Zhang, H.; Evid, J. Rhein: A Review of Pharmacological Activities. *Based Complement. Alternat. Med.* **2015**, 1–10. [CrossRef] [PubMed]
2. Abu, N.; Ali, N.; Ho, W.; Yeap, S.; Aziz, M.Y.; Alitheen, N. Damnacanthal: A Promising Compound as a Medicinal Anthraquinone. *Anticancer Agents Med. Chem.* **2014**, *14*, 750–755. [CrossRef] [PubMed]
3. Zheng, C.J.; Shao, C.L.; Guo, Z.Y.; Chen, J.F.; Deng, D.S.; Yang, K.L.; Chen, Y.Y.; Fu, X.M.; She, Z.G.; Lin, Y.C.; Wang, C.Y. Bioactive hydroanthraquinones and anthraquinone dimers from a soft coral-derived *Alternaria sp.* fungus. *J. Nat. Prod.* **2012**, *75*, 189–197. [CrossRef] [PubMed]
4. Kanamaru, S.; Honma, M.; Murakami, T.; Tsushima, T.; Kudo, S.; Tanaka, K.; Nihei, K.; Nehira, T.; Hashimoto, M. Absolute stereochemistry of altersolanol A and alterporriols. *Chirality.* **2012**, *24*, 137–146. [CrossRef] [PubMed]
5. Suemitsu, R.; Kitagawa, N.; Horie, S.; Kazawa, K.; Harada, T. Isolation and Identification of Altersolanol B (7-Methoxy-2-methyl-(2R, 3S)-1, 2, 3, 4-tetrahydro-2, 3, 5-trihydroxyanthraquinone) from the Mycelium of Alternaría porri (Ellis) Ciferri. *Agric. Biol. Chem.* **2014**, *42*, 1801–1802. [CrossRef]
6. Kasai, M.; Shirahata, K.; Ishii, S.; Mineura, K.; Marumo, H.; Tanaka, H.; OMURA, S. Structure of Nanaomycin E, A New Naaomycin. *J. Antibiot.* **1979**, *5*, 442–445. [CrossRef]
7. Okabe, T.; Nomoto, K.; Tanaka, N. Lactoquinomycin B, A Novel Antibiotic. *J Antibiot.* **1986**, *1*, 1–5. [CrossRef]
8. Becker, A.M.; Rickards, R.W.; Schmalzl, K.J.; Yick, H.C. Metabolites of Dactylaria Lutea, the Structures of Dactylariol and the Antiprotozoal Antibiotic Dactylarin. *J. Antibiot.* **1978**, *4*, 324–329. [CrossRef]
9. Noda, T.; Take, T.; Watanabe, T.; Abe, J. The Structure of Bostrycin. *Tetrahedron* **1970**, *26*, 1339–1346. [CrossRef]
10. Debbab, A.; Aly, A.H.; Edrada-Ebel, R.A.; Wray, V.; Müller, W.E.G.; Totzke, F.; Zirrgiebel, U.; Schächtele, C.; Kubbutat, M.H.G.; Lin, W.H.; et al. Bioactive Metabolites from the Endophytic Fungus *Stemphylium globuliferum* Isolated from Mentha pulegium. *J. Nat. Prod.* **2009**, *72*, 626–631. [CrossRef]
11. Ge, H.M.; Song, Y.C.; Shan, C.Y.; Ye, Y.H.; Tan, R.X. New and cytotoxic anthraquinones from *Pleospora sp.* IFB-E006, an endophytic fungus in Imperata cylindrical. *Planta. Med.* **2005**, *71*, 1063–1065. [CrossRef]
12. Alvi, K.A.; Rabenstein, J. Auxarthrol A and auxarthrol B: two new tetrahydoanthraquinones from Auxarthron umbrinum. *J. Ind. Microbiol. Biotechnol.* **2004**, *31*, 11–15. [CrossRef]
13. Aly, A.H.; Edrada-Ebel, R.; Wray, V.; Müller, W.E.; Kozytska, S.; Hentschel, U.; Proksch, P.; Ebel, R. Bioactive metabolites from the endophytic fungus *Ampelomyces sp.* isolated from the medicinal plant Urospermum picroides. *Phytochemistry* **2008**, *69*, 1716–1725. [CrossRef] [PubMed]
14. Uzor, P.F.; Ebrahim, W.; Osadebe, P.O.; Nwodo, J.N.; Okoye, F.B.; Muller, W.E.; Lin, W.; Liu, Z.; Proksch, P. Metabolites from Combretum dolichopetalum and its associated endophytic fungus Nigrospora oryzae-Evidence for a metabolic partnership. *Fitoterapia* **2015**, *105*, 147–150. [CrossRef]
15. Klaiklay, S.; Rukachaisirikul, V.; Phongpaichit, S.; Pakawatchai, C.; Saithong, S.; Buatong, J.; Preedanon, S.; Sakayaroj, J. Anthraquinone derivatives from the mangrove-derived fungus *Phomopsis sp.* PSU-MA214. *Phytochem. Lett.* **2012**, *5*, 738–742. [CrossRef]
16. Isaka, M.; Chinthanom, P.; Rachtawee, P.; Srichomthong, K.; Srikitikulchai, P.; Kongsaeree, P.; Prabpai, S. Cytotoxic hydroanthraquinones from the mangrove-derived fungus Paradictyoarthrinium diffractum BCC 8704. *J. Antibiot (Tokyo)* **2015**, *68*, 334–338. [CrossRef] [PubMed]
17. Chen, B.; Shen, Q.; Zhu, X.; Lin, Y.C. The Anthraquinone Derivatives from the Fungus Alternaria sp. XZSBG-1 from the Saline Lake in Bange, Tibet, China. *Molecules* **2014**, *19*, 16529–16542. [CrossRef]
18. Kelly, T.R.; Montury, M. A Regiochemically-controlled Synthesis of Altersolanol B. *Tetrahedron. Lett.* **1978**, *45*, 4309–4310. [CrossRef]
19. Locatelli, M. Anthraquinones: Analytical Techniques as a Novel Tool to Investigate on the Triggering of Biological Targets. *Curr. Drug. Targets.* **2011**, *12*, 366–380. [CrossRef]
20. Liu, P.; Wang, P.M.; Xu, J.Z. Assignment of NMR Data of Alterporriol I. *Chem. Nat. Comp.* **2017**, *53*, 653–657.
21. Pretsch, A.; Proksch, P.; Debbab, A. New Anthraquinone Derivatives. U.S. Patent 0 129 927 Al, 24 May 2012.
22. Bai, H.; Kong, W.W.; Shao, C.L.; Li, Y.; Liu, Y.Z.; Liu, M.; Guan, F.F.; Wang, C.Y. Zebrafish Embryo Toxicity Microscale Model for Ichthyotoxicity Evaluation of Marine Natural Products. *Mar. Biotechnol. (NY)* **2016**, *18*, 264–270. [CrossRef] [PubMed]
23. Mosmann, T. Rapid Colorimetric Assay for Cellular Growth and Survival: Application to Proliferation and Cytotoxicity Assays. *J. Immunol. Meth.* **1983**, *65*, 55–63. [CrossRef]

24. Skehan, P.; Storeng, R.; Scudiero, D.; Monks, A.; Mahon, J.M.; Vistica, D.; Warren, J.T.; Bokesch, H.; Kenney, S.; Boyd, M.R. New Colorimetric Cytotoxicity Assay for Anticancer-Drug Screening. *J. Natl. Cancer. Inst.* **1990**, *82*, 1107–1112. [CrossRef]
25. Yu, G.H.; Wu, G.W.; Sun, Z.C.; Zhang, X.M.; Che, Q.; Gu, Q.Q.; Zhu, T.J.; Li, D.H.; Zhang, G.J. Cytotoxic Tetrahydroxanthone Dimers from the Mangrove-Associated Fungus Aspergillus versicolor HDN1009. *Mar. Drugs* **2018**, *16*, 335. [CrossRef] [PubMed]
26. Du, L.; Zhu, T.J.; Liu, H.B.; Fang, Y.C.; Zhu, W.M.; Gu, Q.Q. Cytotoxic polyketides from a marine-derived fungus, *Aspergillus glaucus*. *J. Nat. Prod.* **2008**, *71*, 1837–1842. [CrossRef] [PubMed]
27. Xu, Z.; Liu, R.N.; Guan, H.S. Dual-target inhibitor screening against thrombin and factor Xa simultaneously by mass spectrometry. *Anal. Chim. Acta* **2017**, 1–10. [CrossRef] [PubMed]
28. Tang, J.; Wang, B.X.; Wu, T.; Wan, J.T.; Tu, Z.C.; Njire, M.; Wan, B.J.; Franzblauc, S.G.; Zhang, T.Y.; Lu, X.Y.; et al. Design, Synthesis, and Biological Evaluation of Pyrazolo [1,5-a] pyridine-3-carboxamides as Novel Antitubercular Agents. *ACS Med. Chem. Lett.* **2015**, *6*, 814–818. [CrossRef]

Article

New Diketopiperazines from a Marine-Derived Fungus Strain *Aspergillus versicolor* MF180151

Jiansen Hu [1,2], **Zheng Li** [1,2], **Jieyu Gao** [1,3], **Hongtao He** [1,2], **Huanqin Dai** [1], **Xuekui Xia** [4], **Cuihua Liu** [1,*], **Lixin Zhang** [5,*] and **Fuhang Song** [1,*]

[1] Chinese Academy of Sciences Key Laboratory of Pathogenic Microbiology and Immunology, Institute of Microbiology, Chinese Academy of Sciences, Beijing 100101, China; huzxcv10@126.com (J.H.); bluewave2015@sina.com (Z.L.); jieyu_gao@163.com (J.G.); hehongtao2010@live.cn (H.H.); huanqindai@gmail.com (H.D.)

[2] University of Chinese Academy of Sciences, Beijing 100049, China

[3] School of Food and Biological Engineering, Hefei University of Technology, Hefei 230009, China

[4] Key Biosensor Laboratory of Shandong Provinde, Biology Institute, Qilu University of Technology (Shandong Academy of Sciences), Jinan 250013, China; xiaxk@sdas.org

[5] State Key Laboratory of Bioreactor Engineering, East China University of Science of Technology, Shanghai 200237, China

*   Correspondence: liucuihua@im.ac.cn (C.L.); lxzhang@ecust.edu.cn (L.Z.); songfuhang@im.ac.cn (F.S.); Tel.: +86-10-64806197 (C.L.); +86-21-64252575 (L.Z.); +86-10-64806058 (F.S.)

Received: 3 April 2019; Accepted: 30 April 2019; Published: 2 May 2019

Abstract: Six new diketopiperazines, (±)-7,8-epoxy-brevianamide Q ((±)-**1**), (±)-8-hydroxy-brevianamide R ((±)-**2**), and (±)-8-epihydroxy-brevianamide R ((±)-**3**), together with four known compounds, (±)-brevianamide R ((±)-**4**), versicolorin B (**5**) and averufin (**6**), were isolated from a marine-derived fungus strain *Aspergillus versicolor* MF180151, which was recovered from a sediment sample collected from the Bohai Sea, China. The chemical structures were established by 1D- and 2D-NMR spectra and HR-ESI-MS. **1** is the first sample of brevianamides with an epoxy moiety. Their bioactivities were evaluated against *Candida albicans*, *Bacillus subtilis*, *Staphylococcus aureus*, methicillin-resistant *S. aureus*, *Pseudomonas aeruginosa*, and Bacillus Calmette-Guérin. Compounds **1–4** showed no activities against the pathogens, and compounds **5** and **6** showed moderate activities against *S. aureus* and methicillin-resistant *S. aureus*.

Keywords: marine-derived fungus; *Aspergillus versicolor*; diketopiperazine; antibacterial

---

1. Introduction

Marine-derived fungi are revealed to be excellent resources for novel secondary metabolites and many lead compounds have been characterized for drug development [1–3]. *Aspergillus versicolor*, a slow-growing filamentous fungus, normally are found in air, soil, marine sediment, corrupted plants, and agricultural products. Previous chemical investigations on the fungus *Aspergillus versicolor* from different environments have resulted in the identification of new secondary metabolites with a variety of structures, such as alkaloids [4–9], anthraquinones [10–14], xanthones [15–19], diphenyl ethers [20,21], lactones [22–26], peptides [27], polyketide [28], terpenoids [29,30], and varicuothiols [31].

During our continuous screening of new secondary metabolites from marine *Aspergillus versicolor*, six new diketopiperazines ((±)-**1**–(±)-**3**) named as (±)-7,8-epoxy-brevianamide Q, (±)-8-hydroxy-brevianamide R, and (±)-8-epihydroxy-brevianamide R along with four known compounds (±)-brevianamide R, versicolorin B and averufin ((±)-**4**–**6**, Figure 1) were isolated from the culture material of a marine-derived fungus strain *Aspergillus versicolor* MF180151. Compound **1** is the first sample of brevianamides with an epoxy moiety. In this paper, we describe the fermentation, isolation, structure elucidation and preliminary bioactivities of these compounds.

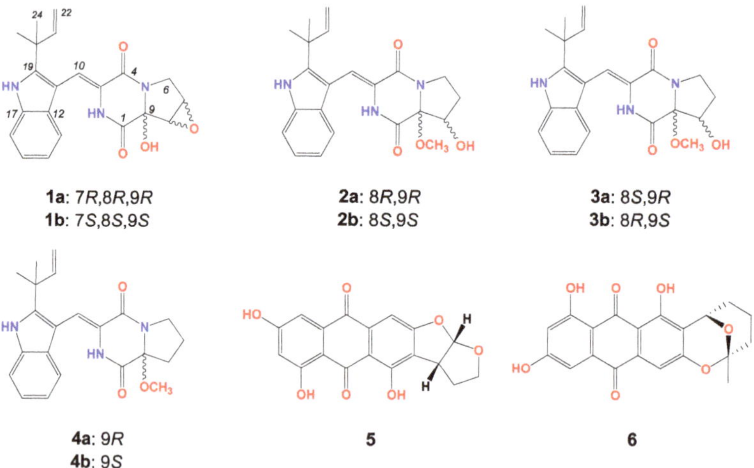

1a: 7*R*,8*R*,9*R*
1b: 7*S*,8*S*,9*S*

2a: 8*R*,9*R*
2b: 8*S*,9*S*

3a: 8*S*,9*R*
3b: 8*R*,9*S*

4a: 9*R*
4b: 9*S*

5

6

Figure 1. The structures of compounds **1–6**.

2. Results

2.1. Characterization and Identification of the Isolated Strain MF180151

The strain MF180151 was isolated from a marine sediment sample from the Bohai Sea, China. The identification of the strain was performed based on the morphology and phylogenetic analysis.

The ITS gene region of ribosomal DNA of the strain was PCR-amplified and sequenced. By comparing the ITS sequence to GenBank, it was indicated that the strain MF180151 belonged to the genus *Aspergillus* and shared a highest similarity with *Aspergillus versicolor* (99.66%). The phylogenetic tree based on ITS gene sequence revealed that the strain MF180151 formed a distinct phylogenetic cluster with *A. versicolor* (Figure 2) with a bootstrap value above 95%.

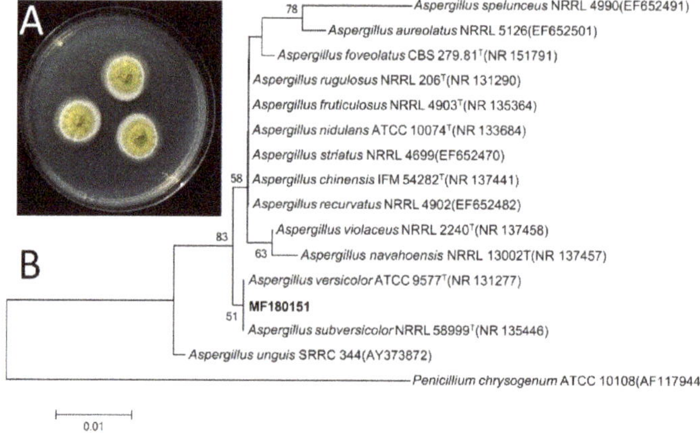

Figure 2. Morphology and neighbor-joining phylogenetic tree of strain MF180151. A: The morphology of the strain MF180151; B: The neighbor-joining phylogenetic tree of strain MF180151, numbers at nodes indicate levels of bootstrap support (%) based on a neighbor-joining analysis of 1,000 resampled datasets; only values >50% are given.

2.2. Structure Elucidation

(±)-7,8-Epoxy-brevianamide Q ((±)-**1**) were isolated as a light yellow amorphous powder. The molecular formula of **1** was established as $C_{21}H_{21}N_3O_4$ by HRESIMS (*m/z* 380.1608 [M + H]$^+$ showed in Figure S7, calcd for $C_{21}H_{22}N_3O_4$: 380.1605). The planar structure of **1** was determined by 1D and 2D NMR spectra analyses, including ^{1}H, ^{13}C, ^{1}H-^{1}H homonuclear correlated spectroscopy (COSY), heteronuclear single quantum coherence (HSQC) and heteronuclear multiple bond correlation (HMBC, Figures S1–S5). The ^{1}H and ^{13}C NMR data of **1** is tabulated in Table 1, which revealed the moieties of indole, diketopiperazine, prenyl, and one isolated double bond. Further analyses of 2D NMR data confirmed these moieties. Additionally, the HMBC correlations from the methyl groups (δ_H 1.50 and 1.45, H$_3$-23 and H$_3$-24) to C-19 (δ_C 144.6), C-20 (δ_C 39.0) and C-21 (δ_C 145.1) suggested that C-20 of the prenyl was attached to C-19 of the indole moiety. The HMBC crossing peaks from H-10 (δ_H 7.02) to C-12 (δ_C 126.2), C-19 (δ_C 144.6) and C-4 (δ_C 160.2) revealed that the diketopiperazine and indole moieties were connected by double bond of C-3 (δ_C 124.5) and C-10 (δ_C 113.4). The connectivity among C-1, C-9 and C-8 was confirmed by the HMBC correlations from 9-OH (δ_H 7.54) to C-1 (δ_C 163.1), C-9 (δ_C 86.0) and C-8 (δ_C 57.4). Thus the planar structure of **1** was assigned as shown in Figure 1. The rotating frame overhauser effect spectroscopy (ROESY, Figure S6) correlation between 2-NH (δ_H 9.37) and H-13 (δ_H 7.32) suggested the *cis* form of the double bond between C-3 and C-10. The ROESY signal from 9-OH (δ_H 7.54) to H-7 (δ_H 3.95)/H-8 (δ_H 3.93) revealed the relative configurations of **1** (Figure 3).

(±)-8-Hydroxy-brevianamide R ((±)-**2**) were isolated as a light yellow amorphous powder. The molecular formula of **2** was established as $C_{22}H_{25}N_3O_4$ by HRESIMS (*m/z* 396.1912 [M + H]$^+$ showed in Figure S15, calcd for $C_{22}H_{26}N_3O_4$: 396.1918). Analyses of the ^{1}H, ^{13}C, COSY and HSQC NMR data (Table 1, Figures S9–S11, S13) revealed that **2** possessed the same carbon skeleton as that of **1**. By comparing the NMR data of **2** with those of brevianamide U [9] and brevianamide R [32], it was revealed that **2** was methylated at the 9-hydroxyl group of brevianamide U, which was confirmed by the HMBC (Figure S12) crossing peak from 9-OMe (δ_H 3.23) to C-9 (δ_C 94.5). The 8-hydroxyl group was confirmed by the HMBC correlations from 8-OH (δ_H 5.52) to C-7 (δ_C 28.4), C-8 (δ_C 74.0) and C-9 (δ_C 94.5). Thus, the planer structure of **2** was assigned. In the ROESY (Figure S14) spectrum of **2**, the correlation between 2-NH (δ_H 9.25) and H-13 (δ_H 7.14) suggested the *cis* form of double bond between C-3 and C-10. And the ROESY correlations between 9-OMe (δ_H 3.23) and H-8 (δ_H 4.22), and the absence between 9-OMe (δ_H 3.23) and 8-OH (δ_H 5.52) indicated that 9-OMe and H-8 were *cis* form. Therefore, the relative configurations of **2** was established as shown in Figure 3.

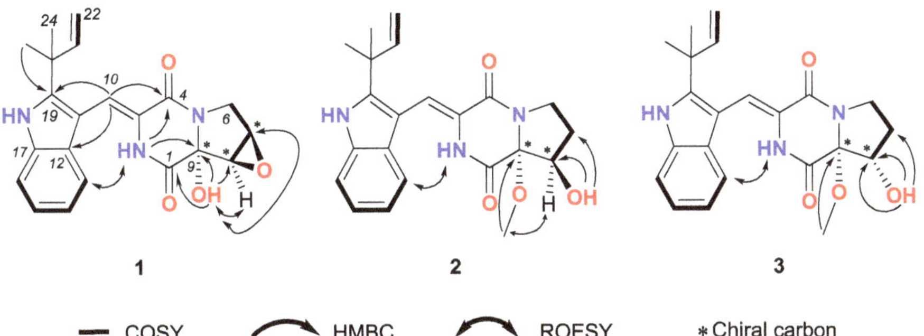

Figure 3. COSY, Key HMBC and ROESY correlations of compounds **1–3**.

Table 1. ^{1}H and ^{13}C NMR data (600 MHz, DMSO-d_6) for compounds **1–3**.

Position	1		2		3	
	δ_C	δ_H, mult (*J* in Hz)	δ_C	δ_H, mult (*J* in Hz)	δ_C	δ_H, mult (*J* in Hz)
1	163.1		160.9		162.9	
2		9.37, s		9.25, s		9.35, s
3	124.5		124.7		124.7	
4	160.2		159.3		160.1	
6a	45.6	3.52, d (13.2)	43.4	3.42, ddd (12.0, 10.2, 1.8)	40.6	3.47, m
6b		3.94, overlap		3.91, ddd (12.0, 8.4, 8.4)		
7a	51.1	3.95, overlap	28.4	2.13, m	27.6	2.12, m
7b				1.76, ddd (13.2, 8.4, 1.8)		1.86, dq (12.0, 9.6)
8	57.4	3.93, overlap	74.0	4.22 dd (4.8, 4.8)	73.6	4.29, ddd (14.4, 6.0, 3.0)
8-OH				5.52, d (4.8)		5.14, d (6.0)
9	86.0		94.5		87.1	
9-OH		7.54, s				
9-OMe			50.6	3.23, s	51.8	3.42, s
10	113.4	7.02, s	112.0	7.04, s	112.6	7.03, s
11	104.0		103.3		103.7	
12	126.2		126.2		126.2	
13	119.7	7.32, d (7.8)	118.6	7.14, d (7.8)	119.0	7.22, d (7.8)
14	119.3	7.00, dd (7.8, 7.8)	119.5	7.02, dd (7.8, 7.8)	119.4	7.02, dd (7.8, 7.8)
15	120.7	7.08, dd (7.8, 7.8)	120.8	7.10, dd (7.8, 7.8)	120.8	7.09, dd (7.8, 7.8)
16	111.4	7.41, d (7.8)	111.7	7.43, d (7.8)	111.6	7.42, d (7.8)
17	135.1		135.2		135.1	
18-NH		11.06, s		11.09, s		11.09, s
19	144.6		144.3		144.5	
20	39.0		39.0		39.0	
21	145.1	6.08, dd (17.4, 10.8)	145.1	6.08, dd (17.4, 10.8)	145.1	6.07, dd (17.4, 10.8)
22a	111.7	5.05, d (17.4)	111.7	5.04, d (17.4)	111.7	5.04, d (17.4)
22b		5.06, d (10.8)		5.06, d (10.8)		5.06, d (10.8)
23	27.4	1.50, s	27.4	1.49, s	27.4	1.49, s
24	27.8	1.45, s	27.7	1.47, s	27.7	1.45, s

(±)-8-Epihydroxy-brevianamide R ((±)-**3**) were isolated as a light yellow amorphous powder. The molecular formula of **3** was established as $C_{22}H_{25}N_3O_4$ by HRESIMS (*m/z* 396.1921 [M+H]$^+$ showed in Figure S22, calcd for $C_{22}H_{26}N_3O_4$: 396.1918). By comparing the ^{1}H, ^{13}C NMR, COSY and HSQC data (Table 1, Figures S17–S19, S21) of **3** with those of **2**, it is revealed that **3** possessed the similar structure as that of **2**. Analyses of the 2D NMR data suggested the same planer structure of **3** and **2**. The 8-hydroxyl group was confirmed by the HMBC (Figure S20) correlations from 8-OH (δ_H 5.14) to C-7 (δ_C 27.6), C-8 (δ_C 73.6) and C-9 (δ_C 87.1). And the HMBC signal from 9-OMe (δ_H 3.42) to C-9 (δ_C 87.1) revealed the methyoxyl group at C-9. In the ROESY (Figure S22) spectrum of **3**, the correlation between 2-NH (δ_H 9.35) and H-13 (δ_H 7.22) suggested the *cis* form of double bond between C-3 and C-10. By comparison the chemical shift of C-9 for **3** with that of **2**, **3** was established as an epimer of **2**, with the relative configuration of 9-OCH$_3$ and 8-OH being *cis* and named as (±)-8-epihydroxy-brevianamide R which was shown in Figure 3.

($\pm$)-**1**–($\pm$)-**3** did not show significant CD spectra absorption (Figure S8, S16, and S23) and optical rotations, $[\alpha]_D^{25}$ +3.00 (*c* 0.1, CH$_3$OH) for ($\pm$)-**1**, $[\alpha]_D^{25}$ +1.00 (*c* 0.1, CH$_3$OH) for ($\pm$)-**2**, and $[\alpha]_D^{25}$ +2.00 (*c* 0.1, CH$_3$OH) for ($\pm$)-**3**, which indicated that ($\pm$)-**1**–($\pm$)-**3** were racemic mixtures.

Additionally, the structures of the three known compounds were identified as ($\pm$)-brevianamide R ((±)-**4**) [32], versicolorin B (**5**) [33] and averufin (**6**) [34] based on its HRESIMS, ^{1}H NMR, and ^{13}C NMR data and comparing with previous reports.

2.3. Biological Activities

The biological activity of those compounds were evaluated against pathogens of Bacillus Calmette-Guérin (BCG), *C. albicans*, *B. subtilis*, *S. aureus*, methicillin-resistant *S. aureus* (MRSA), and *P. aeruginosa*. The new diketopiperazines ($\pm$)-**1**–($\pm$)-**3** and ($\pm$)-brevianamide R ((±)-**4**) showed no significant antibacterial activities against those pathogens. Versicolorin B (**5**) exhibited moderate activities against *S. aureus* and MRSA with the MIC values of 6.25 µg/mL and 12.5 µg/mL. Simultaneously, averufin (**6**) exhibited moderate activities against *S. aureus* and MRSA with the MIC values of 6.25 µg/mL and 25 µg/mL (Table 2).

Table 2. Antimicrobial Activities of **1–6**.

Organism (strain)	Minimum Inhibitory Concentration (µg/mL)						
	1	2	3	4	5	6	Control
Bacillus Calmette–Guérin (Pasteur 1173P2)	>100	>100	>100	>100	>100	>100	0.05 [a]
Staphylococcus aureus (ATCC 6538)	>100	>100	>100	>100	6.25	6.25	1 [b]
methicillin-resistant *S. aureus* (MRSAa)	>100	>100	100	>100	12.5	25	1 [b]
Bacillus subtilis (ATCC 6633)	>100	>100	>100	>100	>100	>100	0.5 [b]
Pseudomonas aeruginosa (PAO1)	>100	>100	>100	>100	>100	>100	1 [c]
Candida albicans (SC 5314)	>100	>100	>100	>100	100	>100	0.016 [d]

[a] Isoniazid; [b] Vancomycin; [c] Ciprofloxacin; [d] Ketoconazole.

3. Discussion

Brevianamides belong to a class of naturally occurring 2,5-diketopiperazine alkaloids, which are mainly produced by fungi of *Penicillium* and *Aspergillus* [6,9,32,35,36]. In this research, three pairs of new brevianamides ((±)-**1**–(±)-**3**) were isolated and their relative configurations were elucidated according to the 1D, 2D NMR, HRESIMS, UV. But, the specific optical rotation analysis and CD showed that these compounds were racemic mixtures. In more than 24 brevianamides, the hydroxy-substitution were mainly occurred at C-8 or/and C-9. In our research, ($\pm$)-7,8-epoxy-brevianamide Q ((±)-**1**) was discovered as the first brevianamide analogues with an epoxy substitution. Compounds ($\pm$)-**1**–($\pm$)-**4** did not exhibit antifungal and antibacterial activities against *C. albicans*, *B. subtilis*, *S. aureus*, MRSA, *P. aeruginosa* and Bacillus Calmette-Guérin (MIC >100 µg/mL). Versicolorin B (**5**) exhibited moderate activities against *S. aureus* and MRSA with the MIC values of 6.25 µg/mL and 12.5 µg/mL. Simultaneously, averufin (**6**) exhibited moderate activities against *S. aureus* and MRSA with the MIC values of 6.25 µg/mL and 25 µg/mL.

4. Materials and Methods

4.1. General Experimental Details

Specific optical rotations ($[\alpha]$D) were measured on AntonPaar MCP 200 polarimeter (Anton Paar GmbH, Graz, Austria) in a 100 × 2 mm cell. CD spectra were measured on Chirascan spectropolarimeter (Applied Photophysics Ltd., Leatherhead, UK) in 1 mm quartz cells. UV-visible spectra were obtained on a Cary 50 spectrophotometer (Varian Inc., Palo Alto, CA, USA) in 1 cm quartz cells. NMR spectra were obtained on a Bruker Avance DRX600 spectrometer (Bruker BioSpin Corp., Billerica, MA, USA) at

600 MHz for ^{1}H and ^{13}C NMR. Chemical shifts were calibrated using residual solvent signals (DMSO-d_6: δ_C 39.5, δ_H 2.50). High-resolution electrospray ionization mass spectrometry measurements were obtained on an Agilent 6520QTOF mass spectrometer (Aglient Technologies Inc., Santa Clara, CA, USA). TLC H silica (Qingdao Marine Chemical Factory, Qingdao, China), Sephadex LH-20 (GE Healthcare BioSciences AB, Uppsala, Sweden) were used for purification. Analytical and semipreparative HPLC was performed using Agilent 1100 or 1200 Series separations modules equipped with Agilent 1100 or 1200 Series diode array detectors and fraction collectors, controlled using ChemStation Rev.B.02.01 (Aglient Technologies Inc., Santa Clara, CA, USA).

4.2. Fungal Culture and Identification

The strain MF180151 was isolated from a marine sediment sample from the Bohai Sea, China. It was incubated on potato dextrose agar (PDA) plate consisting (0.4% potato starch, 2% dextrose, and 2% agar) at 28 °C. The identification was performed based on the morphology and phylogenetic analysis. The whole genomic DNA of the strain was extracted using the E.Z.N.A. kit (Omega Bio-Tek, Norcross, GA, USA). A pair of primers (ITS4: 5"-TCCTCCGCTTATTGATATGC-3"; ITS5: 5"-GGAAGTAAAAGTCGTAACAAGG-3") was used to amplify the ITS region of MF180151. PCR amplification (50.0 µL final volume: 25 µL 2 × Taq Master Mix, 2 µL of 10 µM of each primer, 5.0 µL DNA template and 16 µL ddH$_2$O) of the ITS sequence was performed on Bio-gener PCR Thermal Cycler with the initial denaturation at 95 °C for 3 min, 32 cycles of denaturation (94 °C, 15 s), annealing (60 °C, 15 s), and elongation (72 °C, 60 s), and a final elongation at 72 °C for 5 min. After multiple alignments of ITS sequence of the related species by CLUSTAL W [37], phylogenetic analysis was constructed using neighbor-joining method with bootstrap values based on 1000 replications by MEGA 5.0 [38,39].

The strain was deposited at the Institute of Microbiology, Chinese Academy of Sciences. The nucleotide sequences of ITS gene (accession number MK680178) of *A. versicolor* MF180151 were deposited in GenBank.

4.3. Fermentation, Extraction and Isolation

The strain MF180151 was cultured on potato dextrose agar plate at 28 °C for 7 days. Mature colonies were cut into small pieces (about 1 cm^2) under aseptic conditions. Then, three piece of the strain was inoculated into three 250 mL conical flasks, each containing 40 mL of liquid medium consisting of potato infusion (20%), glucose (2.0%), artificial sea salt (3.5%) and distilled artificial seawater, at 28 °C for 3 d on a rotary shaker at 160 rpm. An aliquot (5 mL) of the resultant seed culture was inoculated into twelve 1 L conical flasks, each containing solid medium consisting of rice (100 g) and artificial seawater (30 mL), and the flasks were incubated stationary for 28 d at 20 °C.

The whole culture media was extracted exhaustively with EtOAc:MeOH (80:20). The combined extracts were reduced to dryness *in vacuo* and the residue was partitioned between EtOAc and H$_2$O. The EtOAc layer (10.3 g) was subjected to a normal phase silica gel chromatography (60 × 80 mm column, TLC H silica) using a stepwise gradient of 50–100% hexane/CH$_2$Cl$_2$ and then 0–100% MeOH/CH$_2$Cl$_2$ to afford 15 fractions (500 mL each). The ninth fraction was chromatographed over a Sephadex LH-20 column (700 × 30 mm) using an isocratic elution of hexane: CH$_2$Cl$_2$:MeOH (5:5:1), to give four subfractions (F1–F4; 100 mL each). Subfraction F3 (205.6 mg) was further purified by HPLC (Agilent Zorbax SB-C18 250 × 9.4 mm, 5 µm column, 3.0 mL/min, isocratic 65% MeCN/H$_2$O) to yield **1** (3.5 mg), **2** (4.2 mg), **3** (5.7 mg), and **4** (4.8 mg). The eighth fraction was purified by HPLC (Agilent Zorbax SB-C18 250 × 9.4 mm, 5 µm column, 3.0 mL/min, isocratic 60% MeCN/H$_2$O) to yield **5** (3.9 mg) and **6** (6.7 mg).

4.4. Biological Activities

The biological activities of isolated compounds were assessed according to the previous report [9]. A panel of human pathogens were used for the assay, including *B. subtilis* (ATCC 6633), *S. aureus*

(ATCC 6538), methicillin-resistant *S. aureus* (clinical strain from Chaoyang Hospital, Beijing, China), and Bacillus Calmette–Guérin (Pasteur 1173P2), *P. aeruginosa* (ATCC 15692), and fungus *C. albicans* (SC 5314).

For general antimicrobial assays, a single colony which was incubated on an LB agar overnight at 37 °C was picked up and suspended in Mueller-Hinton Broth to approximately 1×10^4 cfu/mL. For anti-*C. albicans* assay, a colony of *C. albicans* incubated on a YPD agar plate was picked and suspended in RPMI 1640 to a concentration of 1×10^4 cfu/mL. A twofold serial dilution of each compound to be tested was prepared, and an aliquot of each dilution (2 µL) was added to a 96-well flat-bottom microtiter plate (Greiner). Vancomycin, ciprofloxacin and ketoconazole were used as the positive control and DMSO as the negative control. An aliquot (78 µL) of suspension was then added to each well (to give final compound concentrations of 100 to 0.78 µg/mL in 2.5% DMSO) and the plate was incubated at 37 °C aerobically for 16 h. The MIC was defined as the minimum concentration of the compound that prevented visible growth of the tested bacteria. All the experiments were performed in triplicate.

The strain Bacillus Calmette–Guérin (Pasteur 1173P2) used for the anti-BCG assay was transformed with green fluorescent protein (GFP) constitutive expression plasmid pUV3583c with direct readout of fluorescence as a measure of bacterial growth. The strain was incubated to mid log phase (7 d) at 37 °C in Middle brook 7H9 broth (40 mL; Difco) supplemented with 10% OADC enrichment (Becton Dickinson), 0.05% Tween-80 and 0.2% glycerol and then diluted to an OD_{600} of 0.025 with broth. Aliquots (80 µL) of the bacterial suspension were added to each well of the 96-well micro plates (clear flat-bottom), followed by adding compounds (2 µL in DMSO), which were serially twofold diluted. Isoniazid served as positive control and DMSO as negative control. The plate was incubated at 37 °C for 3 days, and GFP fluorescence was measured with Multi-label Plate Reader using the bottom read mode, with excitation at 485 nm and emission at 535 nm. MIC is defined as the minimum concentration of drug that inhibits more than 90% of bacterial growth reflected by fluorescence value.

Supplementary Materials: The following are available online at http://www.mdpi.com/1660-3397/17/5/262/s1, Figures S1–S8: 1D, 2D NMR, HRESIMS, UV and CD spectra of (±)-7,8-epoxy-brevianamide Q ((±)-**1**), Figures S9–S16: 1D, 2D NMR, HRESIMS, UV and CD spectra of (±)-8-hydroxy-brevianamide R ((±)-**2**), Figures S17–S23: 1D, 2D NMR, HRESIMS, UV and CD spectra of (±)-8-epihydroxy-brevianamide R ((±)-**3**).

Author Contributions: Data curation, J.H. and Z.L.; Investigation, J.H., J.G., H.H., Z.L., H.D. and X.X.; Supervision, F.S., L.Z. and C.L.; Writing—original draft, J.H. and F.S.; review & editing, L.Z., C.L. and F.S.

Funding: This work was supported by grants from the National Key R&D Program of China (2018YFC0311002, 2017YFD0201203, 2017YFC1601300, 2017YFE0108200), the National Natural Science Foundation of China (31600136, 31430002, 31720103901), the Fundamental Research Funds for the Central Public Welfare Research Institutes (Z0547). Taishan Scholarship for Lixin Zhang and the Fundamental Research Funds for the Central Universities 22221818014.

Conflicts of Interest: The authors declare no conflict of interest.

References

1. Blunt, J.W.; Carroll, A.R.; Copp, B.R.; Davis, R.A.; Keyzers, R.A.; Prinsep, M.R. Marine natural products. *Nat. Prod. Rep.* **2018**, *35*, 8–53. [CrossRef] [PubMed]

2. Carroll, A.R.; Copp, B.R.; Davis, R.A.; Keyzers, R.A.; Prinsep, M.R. Marine natural products. *Na. Prod. Rep.* **2019**, *36*, 122–173. [CrossRef] [PubMed]

3. Gomes, N.G.M.; Lefranc, F.; Kijjoa, A.; Kiss, R. Can some marine-derived fungal metabolites become actual anticancer agents? *Mar. Drugs.* **2015**, *13*, 3950–3991. [CrossRef]

4. Li, H.; Sun, W.; Deng, M.; Zhou, Q.; Wang, J.; Liu, J.; Chen, C.; Qi, C.; Luo, Z.; Xue, Y.; et al. Asperversiamides, Linearly fused prenylated indole alkaloids from the marine-derived fungus *Aspergillus versicolor*. *J. Org. Chem.* **2018**, *83*, 8483–8492. [CrossRef]

5. Pan, C.; Shi, Y.; Chen, X.; Chen, C.A.; Tao, X.; Wu, B. New compounds from a hydrothermal vent crab-associated fungus *Aspergillus versicolor* XZ-4. *Org. Biomol. Chem.* **2017**, *15*, 1155–1163. [CrossRef] [PubMed]

6. Peng, J.; Gao, H.; Li, J.; Ai, J.; Geng, M.; Zhang, G.; Zhu, T.; Gu, Q.; Li, D. Prenylated indole diketopiperazines from the marine-derived fungus *Aspergillus versicolor*. *J. Org. Chem.* **2014**, *79*, 7895–7904. [CrossRef]

7. Zhou, M.; Miao, M.M.; Du, G.; Li, X.N.; Shang, S.Z.; Zhao, W.; Liu, Z.H.; Yang, G.Y.; Che, C.T.; Hu, Q.F.; et al. Aspergillines A-E, highly oxygenated hexacyclic indole-tetrahydrofuran-tetramic acid derivatives from *Aspergillus versicolor*. *Org. Lett.* **2014**, *16*, 5016–5019. [CrossRef] [PubMed]

8. Gu, B.B.; Gui, Y.H.; Liu, L.; Su, Z.Y.; Jiao, W.H.; Li, L.; Sun, F.; Wang, S.P.; Yang, F.; Lin, H.W. A new asymmetric diketopiperazine dimer from the sponge-associated fungus *Aspergillus versicolor* 16F-11. *Magn. Reson. Chem.* **2019**, *57*, 49–54. [CrossRef]

9. Song, F.; Liu, X.; Guo, H.; Ren, B.; Chen, C.; Piggott, A.M.; Yu, K.; Gao, H.; Wang, Q.; Liu, M.; et al. Brevianamides with antitubercular potential from a marine-derived isolate of *Aspergillus versicolor*. *Org. Lett.* **2012**, *14*, 4770–4773. [CrossRef] [PubMed]

10. Dou, Y.; Wang, X.; Jiang, D.; Wang, H.; Jiao, Y.; Lou, H.; Wang, X. Metabolites from *Aspergillus versicolor*, an endolichenic fungus from the lichen *Lobaria retigera*. *Drug Discov. Ther.* **2014**, *8*, 84–88. [CrossRef]

11. Hawas, U.W.; El-Beih, A.A.; El-Halawany, A.M. Bioactive anthraquinones from endophytic fungus *Aspergillus versicolor* isolated from red sea algae. *Arch. Pharm. Res.* **2012**, *35*, 1749–1756. [CrossRef]

12. Huang, Z.; Nong, X.; Ren, Z.; Wang, J.; Zhang, X.; Qi, S. Anti-HSV-1, antioxidant and antifouling phenolic compounds from the deep-sea-derived fungus *Aspergillus versicolor* SCSIO 41502. *Bioorg. Med. Chem. Lett.* **2017**, *27*, 787–791. [CrossRef] [PubMed]

13. Wang, W.; Chen, R.; Luo, Z.; Wang, W.; Chen, J. Antimicrobial activity and molecular docking studies of a novel anthraquinone from a marine-derived fungus *Aspergillus versicolor*. *Nat. Prod. Res.* **2018**, *32*, 558–563. [CrossRef]

14. Zhang, Y.; Li, X.M.; Wang, B.G. Anthraquinone derivatives produced by marine-derived fungus *Aspergillus versicolor* EN-7. *Biosci. Biotechnol. Biochem.* **2012**, *76*, 1774–1776. [CrossRef]

15. Song, F.; Ren, B.; Chen, C.; Yu, K.; Liu, X.; Zhang, Y.; Yang, N.; He, H.; Liu, X.; Dai, H.; et al. Three new sterigmatocystin analogues from marine-derived fungus *Aspergillus versicolor* MF359. *Appl. Microbiol. Biotechnol.* **2014**, *98*, 3753–3758. [CrossRef]

16. Wu, G.; Yu, G.; Kurtán, T.; Mándi, A.; Peng, J.; Mo, X.; Liu, M.; Li, H.; Sun, X.; Li, J.; et al. Versixanthones A-F, cytotoxic xanthone-chromanone dimers from the marine-derived fungus *Aspergillus versicolor* HDN1009. *J. Nat. Prod.* **2015**, *78*, 2691–2698. [CrossRef]

17. Li, F.; Guo, W.; Che, Q.; Zhu, T.; Gu, Q.; Li, D. Versicones E-H and arugosin K produced by the mangrove-derived fungus *Aspergillus versicolor* HDN11-84. *J. Antibiot. (Tokyo)*. **2017**, *70*, 174–178. [CrossRef] [PubMed]

18. Wu, Z.H.; Liu, D.; Xu, Y.; Chen, J.L.; Lin, W.H. Antioxidant xanthones and anthraquinones isolated from a marine-derived fungus *Aspergillus versicolor*. *Chin. J. Nat. Med.* **2018**, *16*, 219–224. [CrossRef]

19. Yu, G.; Wu, G.; Sun, Z.; Zhang, X.; Che, Q.; Gu, Q.; Zhu, T.; Li, D.; Zhang, G. Cytotoxic tetrahydroxanthone dimers from the mangrove-associated fungus *Aspergillus versicolor* HDN1009. *Mar. Drugs*. **2018**, *16*, 335. [CrossRef] [PubMed]

20. Gao, H.; Zhou, L.; Cai, S.; Zhang, G.; Zhu, T.; Gu, Q.; Li, D. Diorcinols B-E, new prenylated diphenyl ethers from the marine-derived fungus *Aspergillus versicolor* ZLN-60. *J. Antibiot. (Tokyo)*. **2013**, *66*, 539–542. [CrossRef] [PubMed]

21. Li, X.B.; Zhou, Y.H.; Zhu, R.X.; Chang, W.Q.; Yuan, H.Q.; Gao, W.; Zhang, L.L.; Zhao, Z.T.; Lou, H.X. Identification and biological evaluation of secondary metabolites from the endolichenic fungus *Aspergillus versicolor*. *Chem. Biodivers.* **2015**, *12*, 575–592. [CrossRef] [PubMed]

22. Ding, L.; Xu, P.; Li, T.; Liao, X.; He, S.; Xu, S. Asperfurandiones A and B, two antifungal furandione analogs from a marine-derived fungus *Aspergillus versicolor*. *Nat. Prod. Res.* **2018**, *28*, 1–5. [CrossRef] [PubMed]

23. Li, T.X.; Meng, D.D.; Wang, Y.; An, J.L.; Bai, J.F.; Jia, X.W.; Xu, C.P. Antioxidant coumarin and pyrone derivatives from the insect-associated fungus *Aspergillus versicolor*. *Nat. Prod. Res.* **2018**, *6*, 1–6. [CrossRef]

24. Wang, M.; Sun, M.; Hao, H.; Lu, C. Avertoxins A-D, prenyl asteltoxin derivatives from *Aspergillus versicolor* Y10, an endophytic fungus of *Huperzia serrat*. *J. Nat. Prod.* **2015**, *78*, 3067–3070. [CrossRef]

25. Zhou, M.; Du, G.; Yang, H.Y.; Xia, C.F.; Yang, J.X.; Ye, Y.Q.; Gao, X.M.; Li, X.N.; Hu, Q.F. Antiviral butyrolactones from the endophytic fungus *Aspergillus versicolor*. *Planta. Med.* **2015**, *81*, 235–240. [CrossRef] [PubMed]

26. Zhou, M.; Lou, J.; Li, Y.K.; Wang, Y.D.; Zhou, K.; Ji, B.K.; Dong, W.; Gao, X.M.; Du, G.; Hu, Q.F. Versicolols A and B, two new prenylated isocoumarins from endophytic fungus *Aspergillus versicolor* and their cytotoxic activity. *Arch. Pharm. Res.* **2017**, *40*, 32–36. [CrossRef]

27. Hou, X.M.; Zhang, Y.H.; Hai, Y.; Zheng, J.Y.; Gu, Y.C.; Wang, C.Y.; Shao, C.L. Aspersymmetide A, a new centrosymmetric cyclohexapeptide from the marine-derived Fungus *Aspergillus versicolor*. *Mar. Drugs.* **2017**, *15*, 363. [CrossRef]

28. Salendra, L.; Luo, X.; Lin, X.; Wang, J.; Yang, B.; Zhou, X.; Liu, Y. Versispiroketal A, an unusual tetracyclic bridged spiroketal from the sponge-associated fungus *Aspergillus versicolor* SCSIO 41013. *Org. Biomol. Chem.* **2019**, *17*, 2182–2186. [CrossRef]

29. Guo, Z.Y.; Tan, M.H.; Liu, C.X.; Lv, M.M.; Deng, Z.S.; Cao, F.; Zou, K.; Proksch, P. Aspergoterpenins A−D: four new antimicrobial bisabolane sesquiterpenoid derivatives from an endophytic fungus *Aspergillus versicolor*. *Molecules* **2018**, *23*, 1291. [CrossRef]

30. Li, H.; Sun, W.; Deng, M.; Qi, C.; Chen, C.; Zhu, H.; Luo, Z.; Wang, J.; Xue, Y.; Zhang, Y. Asperversins A and B, two novel meroterpenoids with an unusual 5/6/6/6 ring from the marine-derived fungus *Aspergillus versicolor*. *Mar. Drugs.* **2018**, *16*, 177. [CrossRef]

31. Chen, R.; Liu, D.; Guo, P.; Lin, W. Varicuothiols A and B, new fungal metabolites from *Aspergillus versicolor* with anti-Inflammatory activities. *Chem. Biodivers.* **2018**, *15*, e1700445. [CrossRef] [PubMed]

32. Li, G.Y.; Li, L.M.; Yang, T.; Chen, X.Z.; Fang, D.M.; Zhang, G.L. Four New Alkaloids, Brevianamides O−R, from the Fungus *Aspergillus versicolor*. *Helv. Chim. Acta.* **2010**, *93*, 2075–2080. [CrossRef]

33. Jakšić, D.; Puel, O.; Canlet, C.; Kopjar, N.; Kosalec, I.; Klarić, M.Š. Cytotoxicity and genotoxicity of versicolorins and 5-methoxysterigmatocystin in A549 cells. *Arch. Toxicol.* **2012**, *86*, 1583–1591. [CrossRef]

34. Wu, C.J.; Li, C.W.; Cui, C.B. Seven new and two known lipopeptides as well as five known polyketides: The activated production of silent metabolites in a marine-derived fungus by chemical mutagenesis strategy using diethyl sulphate. *Mar. Drugs.* **2014**, *12*, 1815–1838. [CrossRef] [PubMed]

35. Liu, W.; Wang, L.; Wang, B.; Xu, Y.; Zhu, G.; Lan, M.; Zhu, W.; Sun, K. Diketopiperazine and diphenylether derivatives from marine algae-derived *Aspergillus versicolor* OUCMDZ-2738 by epigenetic activation. *Mar. Drugs.* **2018**, *17*, 6. [CrossRef] [PubMed]

36. Xu, X.; Zhang, X.; Nong, X.; Wang, J.; Qi, S. Brevianamides and mycophenolic acid serivatives from the deep-sea-derived fungus *Penicillium brevicompactum* DFFSCS025. *Mar. Drugs* **2017**, *15*, 43. [CrossRef]

37. Thompson, J.D.; Higgins, D.G.; Gibson, T.J. CLUSTALW: Improving the sensitivity of progressive multiple sequence alignment through sequence weighting, position-specific gap penalties and weight matrix choice. *Nucleic. Acids. Res.* **1994**, *22*, 4673–4680. [CrossRef] [PubMed]

38. Saitou, N.; Nei, M. The neighbor-joining method: A new method for reconstructing phylogenetic trees. *Mol. Biol. Evol.* **1987**, *4*, 406–425. [CrossRef] [PubMed]

39. Tamura, K.; Peterson, D.; Peterson, N.; Stecher, G.; Nei, M.; Kumar, S. MEGA5: Molecular evolutionary genetics analysis using maximum likelihood, evolutionary distance, and maximum parsimony methods. *Mol. Biol. Evol.* **2011**, *28*, 2731–2739. [CrossRef]

 *marine drugs*

Article

Phenalenones from a Marine-Derived Fungus *Penicillium* sp.

Sung Chul Park [1], **Elin Julianti** [1,2], **Sungjin Ahn** [1], **Donghwa Kim** [1], **Sang Kook Lee** [1], **Minsoo Noh** [1], **Dong-Chan Oh** [1], **Ki-Bong Oh** [3,*] and **Jongheon Shin** [1,*]

[1] Natural Products Research Institute, College of Pharmacy, Seoul National University, San 56-1, Sillim, Gwanak, Seoul 151-742, Korea; sungchulpark@snu.ac.kr (S.C.P.); elin_julianti@fa.itb.ac.id (E.J.); sungjinahn@snu.ac.kr (S.A.); dkim0719@snu.ac.kr (D.K.); sklee61@snu.ac.kr (S.K.L.); minsoonoh@snu.ac.kr (M.N.); dongchanoh@snu.ac.kr (D.-C.O.)

[2] School of Pharmacy, Bandung Institute of Technology, Jl. Ganesha 10, Bandung 40132, Indonesia

[3] Department of Agricultural Biotechnology, College of Agriculture and Life Science, Seoul National University, San 56-1, Sillim, Gwanak, Seoul 151-921, Korea

[*] Correspondence: ohkibong@snu.ac.kr (K.-B.O.); shinj@snu.ac.kr (J.S.); Tel.: +82-2-880-4646 (K.-B.O.); +82-2-880-2484 (J.S.)

Received: 26 February 2019; Accepted: 14 March 2019; Published: 18 March 2019

Abstract: Six new phenalenone derivatives (**1–6**), along with five known compounds (**7–11**) of the herqueinone class, were isolated from a marine-derived fungus *Penicillium* sp. The absolute configurations of these compounds were assigned based on chemical modifications and their specific rotations. 4-Hydroxysclerodin (**6**) and an acetone adduct of a triketone (**7**) exhibited moderate anti-angiogenetic and anti-inflammatory activities, respectively, while *ent*-peniciherqueinone (**1**) and isoherqueinone (**9**) exhibited moderate abilities to induce adipogenesis without cytotoxicity.

Keywords: herqueinones; phenalenones; *Penicillium* sp.; marine-derived fungi; adipogenesis; anti-angiogenesis; anti-inflammatory

1. Introduction

Fungi produce a wide variety of polyketide-derived metabolites. Among these, phenalenones are a class of hexa- or heptaketides bearing a perinaphtenone-type tricyclic system [1]. As summarized in a recent comprehensive review, these compounds can have immense structural variations, such as homo- and heterodimerization, the incorporation of additional carbon frameworks, and a high degree of oxygenation and nitrogenation as well as being complexed with metals [2]. A frequently occurring variation is the incorporation of an isoprene unit by forming either a linear ether or a trimethylhydrofuran moiety, and this variation is well-represented in the herqueinones from *Penicillium* sp. [3–5]. Fungi-derived phenalenone compounds have attracted significant interest due to their chemical structures, bioactivities, and biosynthesis [1]. With their diverse phylogenic origins, phenalenones are widely recognized as a representative group of fungal polyketides [1,2].

During the course of our search for novel compounds from marine-derived fungi, we reported the structures of herqueiazole and herqueioxazole, unusual pyrrole- and oxazole-containing phenalenones from a *Penicillium* sp. strain [6]. Herqueidiketal, a cytotoxic sortase A inhibitory congener, also possessed a novel skeleton containing a highly oxidized naphthoquinone moiety [6]. Despite its carbon skeleton being different from typical phenalenones, the presence of naphthalene and dihydrofuran moieties in herqueidiketal may further emphasize the wide structural variations of phenalenones. In our continuing search for such compounds, we isolated several structurally related phenalenones from a large-scale cultivation of this *Penicillium* sp. strain. Here, we report the isolation of eleven compounds (**1–11**) as well as the structure determination of six new compounds (**1–6**) in the herqueinone subclass. 4-Hydroxysclerodin (**6**) exhibited moderate anti-angiogenic activity on human umbilical vascular endothelial cells (HUVECs). The acetone adduct of a triketone (**7**) exhibited moderate anti-inflammatory activity in mouse macrophage RAW 264.7 cells. In addition, *ent*-peniciherqueinone (**1**) and isoherqueinone (**9**) moderately induced adipogenesis in human bone marrow-mesenchymal stem cells (hBM-MSCs). All of these bioactivities were found to occur without cytotoxicity.

2. Results and Discussion

The molecular formula of **1** was deduced to be $C_{20}H_{20}O_8$ with 11 degrees of unsaturation by HRFABMS analysis. The ^{13}C NMR data of this compound showed a signal of a ketone carbon at δ_C 197.6 (Table 1). The signals at δ_C 178.1 and 174.6 could belong to either carbonyl or highly deshielded olefinic carbons. The ^{13}C NMR spectrum, in combination with DEPTs and HSQC spectra (Supplementary Figure S4), displayed nine nonprotonated sp^2 carbon signals in the δ_C 103.5–162.8 region. The deshielded carbons must be one carbonyl carbon and one olefinic carbon, accounting therefore for seven degrees of unsaturation. The ^{13}C NMR data also showed two oxygen-bearing quaternary sp^3 carbons (δ_C 89.5 and 79.0), one methoxy carbon (δ_C 60.9), one shielded quaternary sp^3 carbon (δ_C 46.9), and four shielded methyl carbons (δ_C 16.4, 16.4, 14.9, and 13.3) (Table 2). Combining the NMR data and the degrees of unsaturation, **1** must possess four rings featuring the herqueinone class of phenalenones.

Table 1. ^{13}C NMR (125 and 150 MHz) of **1–6**.

Position	δ_C, Type						
	1 [a]	**1** [b]	**2** [b]	**3** [b]	**4** [b]	**5** [b]	**6** [b]
1	137.3, C	137.1, C	133.5, C	139.0, C	142.7, C	142.9, C	141.8, C
2	103.7, C	103.1, C	102.9, C	103.0, C	116.3, C	116.7, C	115.4, C
3	174.6, C	175.4, C	173.9, C	178.2, C	175.1, C	175.6, C	180.4, C
4	79.0, C	78.4, C	78.2, C	78.5, C	78.5, C	78.5, C	78.9, C
5	197.6, C	198.2, C	198.3, C	197.7, C	193.1, C	193.3, C	192.2, C
6	103.5, C	102.7, C	102.6, C	102.6, C	101.2, C	101.4, C	92.3, C
7	162.8, C	161.7, C	157.2, C	161.9, C	189.9, C	189.3, C	155.5, C
8	131.6, C	131.0, C	129.2, C	131.2, C	76.9, C	75.8, C	
9	161.7, C	161.7, C	156.2, C	163.0, C	200.6, C	200.3, C	164.4, C
10	108.7, C	108.6, C	108.4, C	109.2, C	109.4, C	109.4, C	102.0, C
11	178.1, C	178.7, C	178.7, C	186.4, C	165.6, C	165.8, C	164.0, C
12	143.7, C	144.0, C	143.9, C	122.8, CH	117.5, CH	117.5, CH	117.3, C
13	124.0, C	124.5, C	123.8, C	150.9, C	152.3, C	152.2, C	152.4, C
14	14.9, CH_3	14.6, CH_3	14.4, CH_3	23.8, CH_3	23.5, CH_3	23.4, CH_3	23.2, CH_3
15	60.9, CH_3	60.0, CH_3		60.0, CH_3			
1'	13.3, CH_3	12.9, CH_3	12.9, CH_3	12.9, CH_3	12.7, CH_3	12.8, CH_3	12.8, CH_3
2'	89.5, CH	89.3, CH	88.7, CH	90.6, CH	88.6, CH	88.8, CH	90.8, CH
3'	46.9, C	45.9, C	45.9, C	46.0, C	45.6, C	45.3, C	45.7, C
4'	16.4, CH_3	15.9, CH_3	15.9, CH_3	15.9, CH_3	16.1, CH_3	16.1, CH_3	16.0, CH_3
5'	16.4, CH_3	16.2, CH_3	16.1, CH_3	16.1, CH_3	16.3, CH_3	16.3, CH_3	16.3, CH_3
6'					48.5, CH_2	48.6, CH_2	
7'					206.9, C	207.1, C	
8'					29.8, CH_3	29.7, CH_3	

[a,b] The spectra were recorded in $CDCl_3$ and DMSO-d_6, respectively.

Table 2. ^{1}H NMR (400 and 600 MHz) of **1–6**.

Position	δ_H, Mult. (*J* in Hz)						
	1 [a]	**1** [b]	**2** [b]	**3** [b]	**4** [b]	**5** [b]	**6** [b]
12				6.36, s	6.75, s	6.76, s	6.81, s
14	2.47, s	2.39, s	2.39, s	2.48, s	2.57, s	2.55, s	2.58, s
15	3.92, s	3.77, s		3.77, s			
1′	1.40, d (6.6)	1.35, d (6.3)	1.34, d (6.4)	1.37, d (6.5)	1.30, d (6.4)	1.29, d (6.5)	1.36, d (6.5)
2′	4.99, q (6.6)	4.91, q (6.3)	4.88, q (6.4)	4.99, q (6.5)	4.77, q (6.4)	4.79, q (6.5)	4.92, q (6.5)
4′	1.43, s	1.30, s	1.30, s	0.78, s	0.79, s	0.78, s	0.78, s
5′	0.86, s	0.75, s	0.75, s	1.32, s	1.25, s	1.25, s	1.26, s
6′					3.30, s	3.38, s	
8′					2.09, s	2.12, s	
OH-4	7.23, s	7.38, s	7.28, s	7.52, s	7.15, s	7.23, s	7.41, s
OH-7	13.23, s	13.30, s	13.14, s	13.26, s			
OH-8		8.96, s			6.68, s	6.82, s	
OH-9	13.99, s	14.97, s	14.55, s	15.73, s			
OH-11					12.80, s	12.75, s	11.43, br s
OH-12	6.66, s	9.15, s	9.00, s				

[a,b] The spectra were recorded in CDCl$_3$ and DMSO-d_6, respectively.

Due to the lack of COSY correlations except for that from the methyl doublet at δ_H 1.40 (Me-1′) to the quartet at δ_H 4.99 (H-2′), the structure determination of **1** had to be carried out through extensive HMBC analyses under diverse measuring conditions (Figure 1). First, the long-range couplings from OH-7 (δ_H 13.23) to C-6 (δ_C 103.5), C-7 (δ_C 162.8), and C-8 (δ_C 131.6); from OCH$_3$-8 (δ_H 3.92) to C-8 (δ_C 131.6); and OH-9 (δ_H 13.99) to C-8 (δ_C 131.6), C-9 (δ_C 161.7), and C-10 (δ_C 108.7) lead to a delineation of the C-6 to C-10 fragment. Aided by the four-bond couplings from OH-7 (δ_H 13.23) to C-1 (δ_C 137.3) and OH-9 (δ_H 13.99) to C-1 (δ_C 137.3) by decoupled HMBC (D-HMBC) [7] experiments, the presence of a hexa-substituted benzene ring (C-1, C-6–C-10; ring A) was confirmed. In addition, the combined HMBC and D-HMBC correlations from OH-12 (δ_H 6.66) and H$_3$-14 (δ_H 2.47) to neighboring carbons revealed the presence of an α-hydroxy-β-methyl-α,β-unsaturated ketone group (OH-12 (δ_H 6.66) to C-11 (δ_C 178.1), C-12 (δ_C 143.7), and C-13 (δ_C 124.0); H$_3$-14 (δ_H 2.47) to C-2 (δ_C 103.7), C-11 (δ_C 178.1), C-12 (δ_C 143.7), and C-13 (δ_C 124.0)), which was directly connected to the benzene ring based on the D-HMBC correlation from OH-7 (δ_H 13.23) to C-11 (δ_C 178.1).

In addition to the correlation from H$_3$-14 (δ_H 2.47) to C-2 (δ_C 102.7), a correlation from H$_3$-14 (δ_H 2.47) to a highly deshielded C-3 (δ_C 175.4) in the D-HMBC spectrum was crucial evidence for the attachment of an electron-withdrawing oxygen at this position. Subsequently, long-range correlations from OH-4 (δ_H 7.23) to C-3 (δ_C 8), C4 (δ_C 8), and C5 (δ_C 8) defined not only its connectivity to the C-2 double bond but also placed a carbonyl carbon (δ_C 198.2) at C-5. These carbon–proton correlations constructed an α,β-dioxycyclohexadienone moiety (C-1–C-6; ring C). The assignment of ring C also secured the formation of the conjugated carbonyl group to another six-membered ring (C-1, C-2, C-10–C-13; ring B).

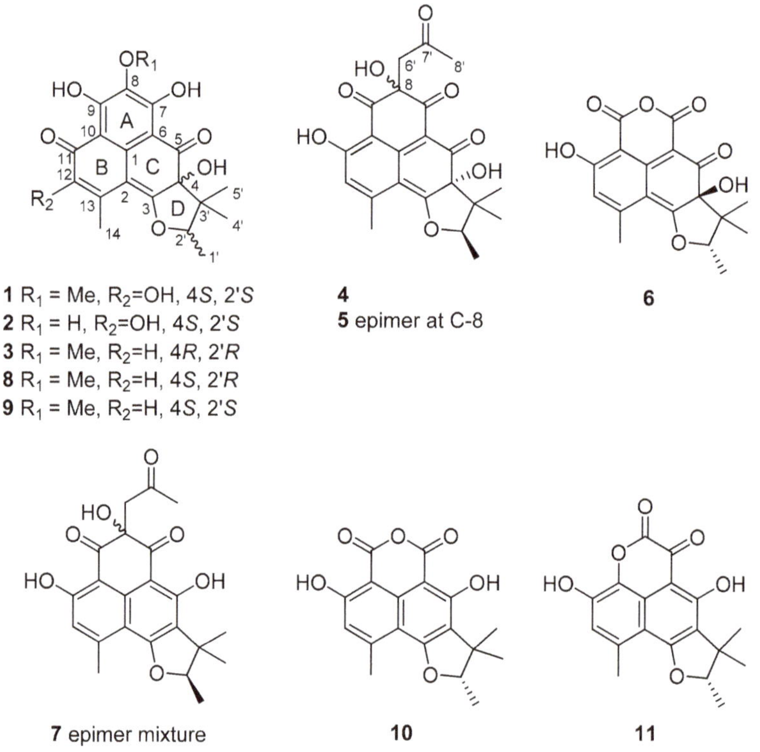

Figure 1. The structures of **1–11**.

The remaining C_5 fragment (C-1′–C-5′) of **1** was readily defined as a 2,3-disubstituted 2-methylbutane moiety by a combination of COSY and HMBC data (Figure 2 and Supplementary Figures S3–S5). The cyclization of this moiety to the three-ring system was also accomplished by a series of long-range carbon–proton correlations. That is, the connection between C-4 and C-5′ was confirmed by the HMBC correlations from H_3-4′ (δ_H 1.43) and H_3-5′ (δ_H 0.86) to C-4 (δ_C 79.0) as well as a long-range correlation from H-2′ to C-5. The diagnostic chemical shifts of the CH-2′ methine group (δ_C 89.3, δ_H 4.91) suggested its attachment to C-3 via an ethereal bridge. This interpretation was corroborated by the correlation from H_3-1′ (δ_H 1.40) to C-3 (δ_C 174.6), which established a hydrofuran moiety (C-3, C-4, C-2′, and C-4′; ring D). Thus, the structure of **1** was defined as a herqueinone-type tetracyclic phenalenone.

The planar structure of **1** was found to be the same as that of the recently reported peniciherqueinone from the fungus *Penicillium herquei* [8]. In our study of the configurations of the C-4 and C-2′ stereogenic centers by NOESY analysis (Figure 3), the OH-4, H-2′, and H_3-5′ protons were oriented toward the same face of the hydrofuran ring based on their mutual cross-peaks. The opposite face was occupied by H_3-1′ and H_3-4′ based on the cross peak between the methyl protons, suggesting that **1** has the same relative configuration (4S* and 2′S*) as peniciherqueinone. Interestingly, despite the same signs of optical rotations, there was a remarkable difference in their values of the specific rotations: $[\alpha]_D^{25}$ (CHCl$_3$) +203 (**1**) and +92 (peniciherqueinone). Since the absolute configurations at C-4 and C-2′ of herqueinones have been the subject of comprehensive investigations [9,10], the discrepancy in the specific rotations of **1** and herqueinones needed to be justified. Using a pre-established chemical modification technique [11–13], **1** was reduced to **1a**, which showed a negative specific rotation ($[\alpha]_D^{25}$ (CHCl$_3$) −23); thus, the 2′S configuration was confirmed. The absolute configuration was further evaluated via the acetylation of **1a** to corresponding 9,11,12-triacetyl derivative **1b** (Figure 4). The sign of the specific rotation of **1b** ($[\alpha]_D^{25}$ (MeOH) -42) was opposite to that of herqueinone (**8**) but the same as that of isoherqueinone (**9**), which proved a $2'S$ configuration [9,10]. Therefore, the absolute configuration of **1** was assigned as $4S$ and $2'S$. Thus, **1**, designated as *ent*-peniciherqueinone, is a new herqueinone-type phenalenone.

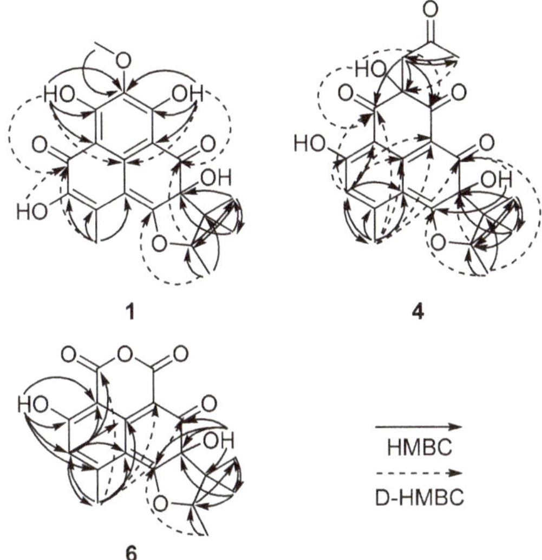

Figure 2. Key correlations of HMBC (arrows) and decoupled HMBC (D-HMBC) (dashed arrows) of **1**, **4**, and **6**.

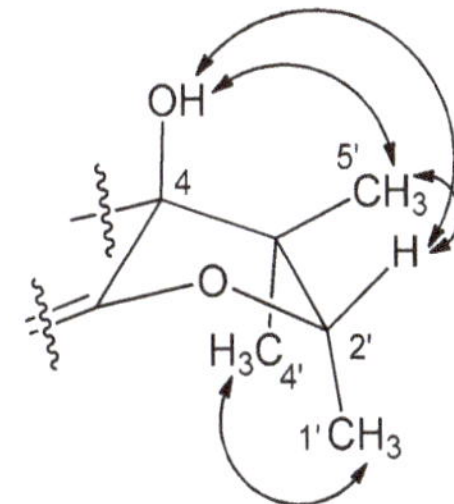

Figure 3. NOESY correlations of the hydrofuran moiety of **1**.

The molecular formula of **2** was deduced to be $C_{19}H_{18}O_8$ based on HRFABMS analysis. The NMR data of this compound were very similar to those of **1**, with the absence of a methyl group. A detailed examination of the ^{13}C and ^1H NMR data revealed that the OMe-8 of **1** (δ_C 60.9, δ_H 3.92) was replaced by a hydroxyl group (δ_H 8.96) in **2**, and this assignment was confirmed by a combination of two-dimensional (2D) NMR analyses. The NOESY data and specific rotation of the reduction product **2a** indicated the same $4S$ and $2'S$ configuration as in **1**. Thus, **2**, designated as 12-hydroxynorherqueinone, was determined to be 8-demethyl-*ent*-peniciherqueinone.

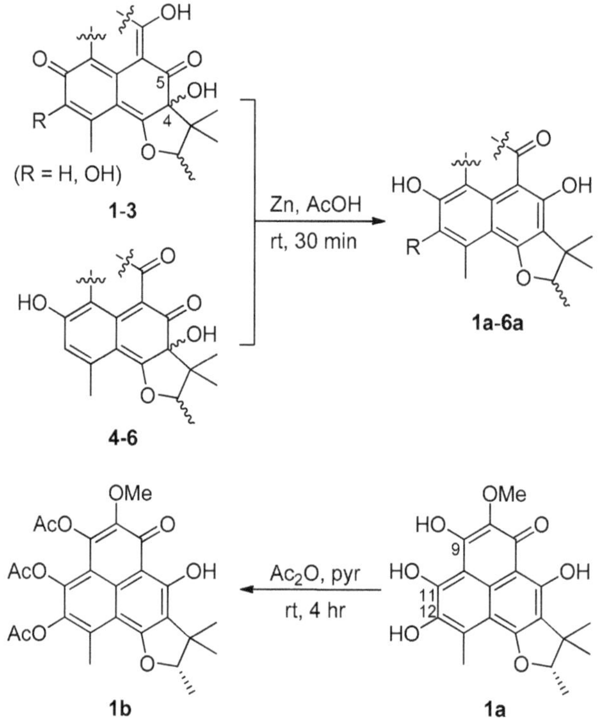

Figure 4. Phenolic derivatization of herqueinones.

Compound **3** was isolated as an orange amorphous solid with a molecular formula of $C_{20}H_{20}O_7$, based on HRFABMS analysis. The ^{13}C and ^1H NMR data of this compound were similar to those obtained for **1**. The most noticeable difference was the replacement of a hydroxyl-bearing olefinic carbon with the sp^2 methine carbons (δ_C 122.8, δ_H 6.36). The structural difference was found to be at C-12 based on the HMBC correlations from H-12 (δ_H 6.36) to C-2 (δ_C 103.0), C-10 (δ_C 109.2), and C-14 (δ_C 23.8) as well as from H_3-14 (δ_H 2.48) to C-2 (δ_C 103.0), C-12 (δ_C 122.8), and C-13 (δ_C 150.9). However, the sign of the specific rotation of **3** ($[\alpha]_D^{25}$ (MeOH) -69) was opposite to those of **1** and **2**, implying a configurational difference. Since the NOESY spectrum showed the same cross-peaks for the hydrofuran moiety as those in the congeners, **3** was proposed to possess the opposite absolute configuration at C-4 and C-2′. As the reduction product of **3** (**3a**) is dextrorotatory (specific rotation ($[\alpha]_D^{25}$ (MeOH) $+39$)), the configuration of C-4 and C-2′ are 4*R*, 2′*R*, respectively. Thus, **3**, designated as *ent*-isoherqueinone, is a new herqueinone-type phenalenone derivative.

The molecular formula of **4** was also established as $C_{22}H_{22}O_8$ by HRFABMS analysis. Although its spectroscopic data resembled those of **1–3**, several differences were found in both ^{13}C and ^1H NMR data. First, aided by the HSQC data, it was found that three additional carbons, i.e., one carbonyl (δ_C 206.9), one methylene sp^2 (δ_C 48.5, δ_H 3.30), and one methyl (δ_C 29.8, δ_H 2.09), were present in this compound (Tables 1 and 2). In the ^{13}C NMR spectrum, resonances of three ketone groups (δ_C 200.6, 193.1, and 189.9) were found for **4**, unlike **1–3**. In addition, an aromatic or olefinic carbon had been replaced by an oxygen-bearing nonprotonated sp^3 carbon (δ_C 76.9). A detailed examination of its NMR data revealed that **4** contained the same B and D rings as **1–3**, and the structural differences were located in the remaining portion of the molecule.

The planar structure of **4** was established by extensive HMBC experiments (Figure 2). Several HMBC correlations were found from an aromatic proton (δ_H 6.75, H-12) and a benzylic methyl proton

(δ_H 2.57, H$_3$-14) to their neighboring carbons (H-12 (δ_H 6.75) to C-2 (δ_C 116.3), C-10 (δ_C 109.4), and C-14 (δ_C 23.5); H$_3$-14 (δ_H 2.57) to C-2 (δ_C 116.3), C-12 (δ_C 117.5), and C-13 (δ_C 152.3)). Aided by the D-HMBC correlations from H-12 (δ_H 6.75) to C-1 (δ_C 142.7) and C-11 (δ_C 165.6) and from H$_3$-14 (δ_H 2.57) to C-1 (δ_C 142.7), the long-range carbon–proton correlations led to the establishment of a hydroxyl- and methyl-bearing pentasubstituted benzene as ring B. Additional D-HMBC correlations from these protons to the conspicuous H$_3$-1' at δ_H 1.30 (H-12 (δ_H 6.75) to C-6 (δ_C 101.2); H$_3$-14 (δ_H 2.57) to C-3 (δ_C 175.1), C-5 (δ_C 193.1), and C-6 (δ_C 101.2); H$_3$-1' (δ_H 1.30) to C-3 (δ_C 175.1) and C-5 (δ_C 193.1)) defined ring C as a hydroxyl-bearing cyclohexadienone. The ring D was found to be the same as that in other herqueinones by a 2D NMR spectrum.

The remaining portion of **4** consists of three ketone carbonyl (δ_C 206.9, 200.6, and 189.9) and one nonprotonated sp^3 (δ_C 76.9), one methylene sp^3 (δ_C 48.5), and one methyl (δ_C 29.8) carbons. These carbons were initially assembled into a 2-keto-propyl group (C-6'–C-8') by the HMBC correlations from the methylene and methyl protons to their neighboring carbons (H$_2$-6' (δ_H 3.30) to C-7' (δ_C 206.9) and C-8' (δ_C 29.8); H$_3$-8' (δ_H 2.09) to C-6' (δ_C 48.5), and C-7' (δ_C 206.9)) (Figure 2). Then, this fragment was connected to the core structure by the HMBC correlations from H$_2$-6' (H$_2$-6' (δ_H 3.30) to C-7 (δ_C 189.9), C-8 (δ_C 76.9), and C-9 (δ_C 200.6)). The confirmation of this assignment as well as the linkage with the B-ring was accomplished by the key D-HMBC correlations from OH-8 (δ_H 6.68) to C-7 (δ_C 189.9) and C-9 (δ_C 200.6); from H-12 (δ_H 6.75) to C-9 (δ_C 200.6); from H$_3$-14 (δ_H 2.57) to C-9 (δ_C 200.6); and from H$_3$-8' (δ_H 2.09) to C-8 (δ_C 76.9). Although it could not be confirmed by 2D-NMR-based carbon–proton correlations, the presence of the four rings, required by the molecular formula and NMR data, directly connected C-6 and C-7 carbonyl carbons to be part of a diketo-bearing six-membered ring as ring A. Thus, the structure of **4** was determined to be a phenalenone related to an acetone adduct of a triketone [14,15].

The molecular formula of **5** was the same as that of **4**, $C_{22}H_{22}O_8$. Moreover, the ^{13}C and ^{1}H NMR data of these compounds were very similar (Tables 1 and 2). Two-dimensional NMR analyses showed the same carbon–proton correlations throughout the entire molecule, indicating that they have the same planar structure. Therefore, **5** could be an epimer of **4**.

In order to clarify the difference in stereochemistry between **4** and **5**, NOESY experiments were carried out. The NOESY spectra of both compounds showed the same cross-peaks around the D ring as those observed in other herqueinones, suggesting the 4R,2'R or 4S,2'S configurations. Then, by chemical conversions to remove the other two stereogenic centers, the absolute configurations at C-4 were determined. That is, **4** and **5** were reduced to **4a** and **5a**, respectively (Figure 4), then the compounds were dehydrated to yield 8,15-unsaturated derivatives **4b** and **5b**, respectively (Figure 5), and the MS and NMR data of these compounds were identical. Furthermore, their specific rotations were also very similar ($[\alpha]_D^{25}$ (CHCl$_3$) -27 and -26 for **4b** and **5b**, respectively), implying that these were indeed the same compound. The negative specific rotations allow us to confidently assign the 2'R configuration for both natural products. Thus, **4** and **5**, designated as oxopropylisoherqueinones A and B respectively, were elucidated as new phenalenones possessing C$_3$ side chains. These compounds possessed 4R, 2'R configurations. However, the configurations at C-8 remain unassigned despite various chemical and spectroscopic analyses.

In order to determine the absolute configurations at C-8 of **4** and **5**, a comparison of the experimental and calculated ECD spectra was carried out. Initially, the experimental CD profiles of these compounds showed opposite signs in the region of 285–340 nm, possibly reflecting the different configuration at C-8 (Supplementary Figure S51). Despite all the efforts, however, the calculated ECD profiles based on the postulated conformational populations failed to assign the absolute configurations satisfactorily (Supplementary Figure S52). This could be due to a weak contribution of a single and remote stereogenic center to the ECD in the molecule possessing several UV chromophores and stereogenic centers.

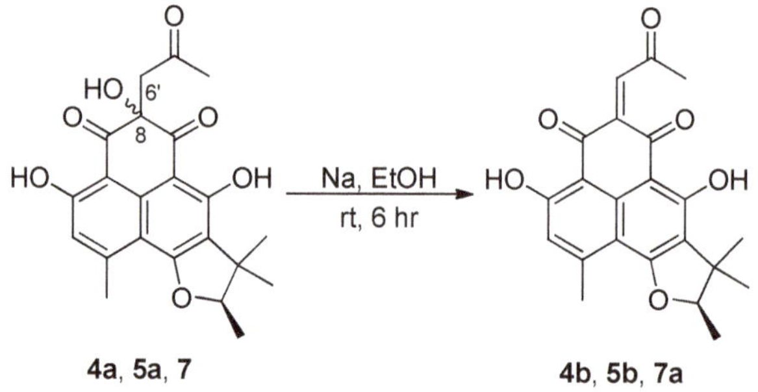

Figure 5. Dehydrations of **4a**, **5a**, and **7**.

The molecular formula of **6** was deduced to be $C_{18}H_{16}O_7$, which corresponds to 11 degrees of unsaturation, by HRFABMS analysis. The ^{13}C and ^1H NMR data of this compound revealed that it is a phenalenone derivative based on the presence of signals for two aromatic rings and a trimethylhydrofuran moiety, which account for eight degrees of unsaturation (Tables 1 and 2). However, only the carbon signals of two nonprotonated quaternary sp^2 carbons (δ_C 164.4 and 155.4) had replaced the NMR signals of the A ring of the other compounds. Therefore, in addition to satisfying the three remaining degrees of unsaturation, the C_2O_3 portion must account for two carbonyls and a cyclic ether or ester group.

The planar structure of **6** was determined with the aid of HMBC experiments (Figure 2). First, the long-range couplings of key protons, such as the four methyl groups (H$_3$-14 (δ_H 2.58), H$_3$-1' (δ_H 1.36), H$_3$-4' (δ_H 0.78), and H$_3$-5' (δ_H 1.26)), an aromatic proton (H-12 (δ_H 6.81)), and two hydroxy protons (OH-4 (δ_H 7.41) and OH-11 (δ_H 11.43)), with their neighboring carbons confirmed the presence of the same B-D polycyclic moiety as in **4** and **5**. An additional coupling to OH-11 placed a carbonyl carbon (δ_C 164.4) at C-9, which was supported by the key D-HMBC correlation from H$_3$-14 (δ_H 2.58) to C-9 (δ_C 164.4). The other carbon (δ_C 155.5) must be located at C-7 due to the shielding of C-6 (δ_C 92.3). Although it was not directly proved by NMR spectra, both the MS data and the shielded chemical shifts of the C-7 and C-9 carbonyls were indicative of an oxygen bridge between these positions, leading to a six-membered cyclic acid anhydride moiety as ring A. The NMR data of the ring portion of **6** were similar to those of sclerodin (**10**), which was previously reported from the fungus *Gremmeniella abietina* thus supporting the structure of **6** [14,16].

The NOESY correlations of **6** placed the OH-4, H-2', and H$_3$-5' on one side and H$_3$-1' and H$_3$-4' on the other side of the hydrofuran moiety, leading to the same relative configuration (4S* and 2'S*) as that in **1–3**. Then, the specific rotation of **6** was similar to that of **3** ($[\alpha]_D^{25}$ −69 and −52 for **3** and **6**, respectively), suggesting they have the same absolute configuration (4R and 2'R). However, to remove the effect of structural differences in ring A, the reduction of **6** produced the 4-deoxy derivative **6a** (=**10**), which possessed only the C-2' stereogenic center (Figure 4). Interestingly, the specific rotation of **6a** showed the same sign as those of **1a** and **2a** but opposite to those of **3a** ($[\alpha]_D^{25}$ +34 and −18 for **3a** and **6a**, respectively), confirming the 2'S configuration. Our results were in good agreement with the specific rotations of natural **6a** (**10**) and 2'-*epi*-**6a**, which are levorotatory and dextrorotatory, respectively [14]. Overall, the configuration of this compound was assigned as 4S,2'S. Notably, changing the phenolic A ring to an acid anhydride inverted the sign of the specific rotation of the herqueinone. Thus, **6**, designated as 4-hydroxysclerodin, is a new phenalenone derivative and structurally related to sclerodin (**10**).

In addition to **1–6**, five previously reported phenalenones (**7–11**) were also isolated. Based on a combination of spectroscopic analyses and a literature survey, these compounds were identified as an acetone adduct of the triketone (**7**) [14], herqueinone (**8**) [3,17,18], isoherqueinone (**9**) [19,20], sclerodin (**10**) [14], and scleroderolide (**11**) [21]. The NMR data of these compounds were in good agreement with the reported values in the literature. Compound **7** was obtained as an unseparated epimeric mixture, which was consistent with the literature [14,15]. Compound **7** was dehydrated to **7a** by the same method used for **4** and **5**, and the $2'R$ configuration was thus assigned. In this way, the epimerization of **7** was found to occur not at C-2′ (in the hydrofuran moiety) but at the hydroxy-bearing C-8 stereogenic center.

Compounds **4**, **5**, and **7** possessed a C_3 oxopropyl moiety (C-6′–C-8′) whose structural resemblance raised the hypothesis that **7** could be the acetone adduct formed during the separation process. This hypothesis has a reliable experimental basis of chemical transformation of a triketone to **7** [14]. In order to verify if **7** is an acetone adduct or a true natural compound biosynthesized by the fungus, the production of these compounds was monitored by time-scale cultivation and LC-ESI-MS analysis. Weekly mass analysis of the culture media showed that the major metabolite **7** was clearly detected after 6 weeks without using acetone (Supplementary Figure S53). Thus, these compounds were unambiguously proved to be the natural products produced by the *Penicillum* sp. fungus.

Although fungal phenalenones exhibit diverse bioactivities [1,2], herqueinone-type compounds have not frequently shown remarkable bioactivities. The mild antioxidant and radical scavenging activities of isoherqueinone (**9**) [9], the antibacterial activity of scleroderolide (**11**) [22], and human leukocyte elastase inhibition of atrovenetinone can be considered exceptions [23]. Regarding the bioactivities of herqueinones, it is interesting to note that the presence of both OH-5 and OH-11 groups are required for the antibacterial activity [1]. The cytotoxicity assay revealed that **1–11** were inactive (IC_{50} > 10 μM) against the K562 (human chronic myeloid leukemia) and A549 (adenocarcinomic human alveolar basal epithelial) cancer cell lines. These compounds were also inactive (MIC > 128 μM) against various bacterial and fungal strains, which was consistent with the report on the structure-activity relationships of herqueinones [1].

Compound **7** moderately inhibited NO production in RAW 264.7 cells with an IC_{50} value of 3.2 μM, while the rest of the isolated compounds were inactive (IC_{50} > 20 μM). In the angiogenesis assay, **6** inhibited tube formation in HUVECs with an IC_{50} of 20.9 μM (Supplementary Table S1 and Figures S54 and S55), while **1** and **9** induced adipogenesis through PPARγ binding and adiponectin secretion-promoting activity in hBM-MSCs and in a concentration-dependent manner, which was determined by adiponectin secretion-promoting effects with their IC_{50} values of 57.5 μM and 39.7 μM, respectively (Table S1 and Figure S56). All of these bioactivities were found to occur without significant cytotoxicity.

In summary, 11 polyketide-derived phenalenones, including six previously unreported phenalenones, were isolated from the culture broth of a marine-derived *Penicillium* sp. The absolute configurations of the stereogenic centers in the hydrofuran ring were assigned by chemical modifications and measurements of specific rotations. Compounds **1**, **6**, **7**, and **9** exhibited diverse bioactivities, such as anti-inflammatory, anti-angiogenetic, and adipogenesis-inducing activities.

3. Materials and Methods

3.1. General Experimental Procedures

Optical rotations were measured on a JASCO P-1020 polarimeter (Easton, MD, USA) using a cell with a 1-cm path length. UV spectra were acquired using a Hitachi U-3010 spectrophotometer (Tokyo, Japan). CD spectra were recorded on an Applied Photophysics Ltd. Chirascan plus CD spectrometer (Applied Photophysics Ltd., Leatherhead, Surrey, UK). IR spectra were recorded on a JASCO 4200 FT-IR spectrometer (Easton, MD, USA) using a ZnSe cell. NMR spectra were recorded in DMSO-d_6 or CDCl$_3$ solutions on Bruker Avance-400, -500, or -600 instruments (Billerica, MA, USA). High-resolution FABMS data were acquired using a JEOL JMS 700 mass spectrometer (Tokyo, Japan) with 6 keV-energy, emission current 5.0 mA, xenon as inert gas, and meta-nitrobenzyl alcohol (NBA) as the matrix at the Korea Basic Science Institute (Daegu, Korea). Low-resolution ESIMS data were recorded on an Agilent Technologies 6130 quadrupole mass spectrometer (Santa Clara, CA, USA) with an Agilent Technologies 1200 series HPLC (Santa Clara, CA, USA). HPLC separations were performed on a SpectraSYSTEM p2000 equipped with a refractive index detector (SpectraSYSTEM RI-150 (Waltham, MA, USA)) and a UV-Vis detector (Gilson UV-Vis-151 (Middleton, WI, USA)). All solvents used were of spectroscopic grade or were distilled prior to use.

3.2. Fungal Material

The fungal strain *Penicillium* sp. was isolated from marine sediments collected from Gagudo, Korea, in October 2008. The isolate was identified using standard molecular biological protocols by DNA amplification and sequencing of the ITS region. Genomic DNA extraction was performed using Intron's i-genomic BYF DNA Extraction Mini Kit according to the manufacturer's protocol. The nucleotide sequence was deposited in the GenBank database under the accession number JF901804. The 18S rDNA sequence of this strain showed 99% identity with *Penicillium herquei* GA4 (GenBank accession number EF536027).

3.3. Extraction and Isolation

The fungus was cultivated on YPG medium (5 g of yeast extract, 5 g of peptone, 10 g of glucose in 1 L of artificial seawater) in 2.8 L Fernbach flasks at 30 °C under static conditions in the dark for 6 weeks. The mycelia and culture broth were separated by filtration, and the broth (20 L) was extracted with EtOAc (20 L × 3). The solvent was evaporated under reduced pressure to afford a crude EtOAc extract (6.2 g), which was fractionated by C18 reversed-phase vacuum flash chromatography using mixtures of H_2O-MeOH, from 50:50 to 0:100, and acetone and EtOAc as the eluents.

Based on the ^1H NMR and LC-MS analyses, the moderately polar fractions (30:70–10:90 H_2O-MeOH) were chosen for further separation. The fraction (220 mg) that eluted with H_2O-MeOH (30:70) was separated by a semi-preparative reversed-phase HPLC (YMC-ODS-A column, 10 × 250 mm; H_2O-MeOH, 45:55; 1.7 mL/min) to yield **4** (t_R = 18.4 min, 5.5 mg) and **5** (t_R = 18.9 min, 7.7 mg). The fraction (570 mg) that eluted with H_2O-MeOH (20:80) was separated by a semi-preparative reversed-phase HPLC (H_2O-MeOH, 32:68; 1.7 mL/min) to afford **1** (t_R = 37.5 min), **2** (t_R = 27.8 min), **3** (t_R = 29.1 min), **8** (t_R = 25.1 min), and **9** (t_R = 21.8 min). Compounds **1** (311.5 mg), **3** (5.6 mg), **8** (16.5 mg), and **9** (4.4 mg) were purified by an analytical HPLC (YMC-ODS-A column, 4.6 × 250 mm; H_2O-MeOH, 37:63; 0.7 mL/min; t_R = 38.8, 34.5, 30.9, and 27.1 min, respectively). Compound **2** (1.7 mg) was also purified by an analytical HPLC (H_2O-MeCN, 48:52; 0.7 mL/min; t_R = 35.0 min). The fraction (230 mg) eluted with H_2O-MeOH (10:90) was separated by a semi-preparative reversed-phase HPLC (H_2O-MeOH, 22:78; 1.7 mL/min) to yield **7** (t_R = 19.7 min, 73.4 mg), **10** (t_R = 22.8 min), and **11** (t_R = 23.5 min). Compounds **10** (3.9 mg) and **11** (3.3 mg) were further purified by an analytical HPLC (H_2O-MeOH, 26:74; 0.7 mL/min; t_R = 26.8 and 30.1 min, respectively).

ent-Peniciherqueinone (**1**): red, amorphous solid; $[\alpha]_D^{25}$ +203 (*c* 1.7, CHCl$_3$), +254 (*c* 1.0, MeOH); UV (MeOH) λ_{max} (log ε) 217 (4.32), 248 (4.27), 311 (4.20), 427 (3.75) nm; IR (ZnSe) ν_{max} 3413 (br), 1629, 1590, 1385 cm^{-1}; ^1H and ^{13}C NMR data, Tables 1 and 2; HRFABMS *m/z* 389.1239 [M + H]$^+$ (calcd for C$_{20}$H$_{21}$O$_8$, 389.1239).

12-Hydroxynorherqueinone (**2**): red, amorphous solid; $[\alpha]_D^{25}$ +124 (*c* 0.1, MeOH); UV (MeOH) λ_{max} (log ε) 217 (4.32), 248 (4.31), 311 (4.36), 430 (3.80) nm; IR (ZnSe) ν_{max} 3445 (br), 1629, 1579, 1461 cm^{-1}; ^1H and ^{13}C NMR data, Tables 1 and 2; HRFABMS *m/z* 375.1079 [M + H]$^+$ (calcd for C$_{19}$H$_{19}$O$_8$, 375.1080).

ent-Isoherqueinone (**3**): orange, amorphous solid; $[\alpha]_D^{25}$ −69 (*c* 0.2, MeOH); UV (MeOH) λ_{max} (log ε) 217 (4.32), 248 (4.29), 311 (4.22), 428 (3.79) nm; IR (ZnSe) ν_{max} 3422 (br), 1631, 1460 cm^{-1}; ^1H and ^{13}C NMR data, Tables 1 and 2; HRFABMS *m/z* 373.1285 [M + H]$^+$ (calcd for C$_{20}$H$_{21}$O$_7$, 373.1283).

Oxopropylisoherquelone A (**4**): brown, amorphous solid; $[\alpha]_D^{25}$ +92 (*c* 0.2, MeOH); UV (MeOH) λ_{max} (log ε) 224 (4.36), 274 (4.30), 357 (3.57) nm; IR (ZnSe) ν_{max} 3382 (br), 1678, 1639, 1297 cm^{-1}; ^{1}H and ^{13}C NMR data, Tables 1 and 2; HRFABMS *m/z* 415.1396 [M + H]$^+$ (calcd for $C_{22}H_{23}O_8$, 415.1393).

Oxopropylisoherquelone B (**5**): brown, amorphous solid; $[\alpha]_D^{25}$ +43 (*c* 0.2, MeOH); UV (MeOH) λ_{max} (log ε) 224 (4.36), 274 (4.30), 357 (3.57) nm; IR (ZnSe) ν_{max} 3415 (br), 1679, 1640, 1297 cm^{-1}; ^{1}H and ^{13}C NMR data, Tables 1 and 2; HRFABMS *m/z* 415.1396 [M + H]$^+$ (calcd for $C_{22}H_{23}O_8$, 415.1393).

4-Hydroxysclerodin (**6**): yellow, amorphous solid; $[\alpha]_D^{25}$ −52 (*c* 0.2, MeOH); UV (MeOH) λ_{max} (log ε) 213 (3.94), 280 (4.21), 312 (3.68) nm; IR (ZnSe) ν_{max} 3424 (br), 3069, 1729, 1460, 1286 cm^{-1}; ^{1}H and ^{13}C NMR data, Tables 1 and 2; HRFABMS *m/z* 345.0977 [M + H]$^+$ (calcd for $C_{18}H_{17}O_7$, 345.0974).

3.4. Reduction of Herqueinones (**1–6**)

To a solution of 44.3 mg (114 µM) of **1** in 0.5 mL of glacial acetic acid was added 100.0 mg (1.53 mM) of zinc dust under nitrogen atmosphere. The mixture was stirred at room temperature for 30 min and filtered through cotton with 1.0 mL of distilled water. The filtrate was left to stand for 45 min and extracted with 1.5 mL of ethyl acetate. Purification by analytical HPLC (YMC-ODS-A column, 4.6 × 250 mm; H$_2$O-MeCN (50:50); 0.7 mL/min) afforded the 4-deoxy derivative (**1a**, 6.8 mg) (t_R = 15.8 min) as a pure compound. Compounds **2–6** were reduced in a similar manner.

4-Deoxy-*ent*-peniciherqueinone (**1a**): $[\alpha]_D^{25}$ −23 (*c* 0.5, CHCl$_3$); ^{1}H NMR (CDCl$_3$, 400 MHz) δ_H 13.14 (1H, s), 13.10 (1H, s), 13.07 (1H, s), 4.57 (1H, q, *J* = 6.5 Hz), 3.99 (3H, s), 2.71 (3H, s), 1.50 (3H, s), 1.44 (3H, d, *J* = 6.5 Hz), 1.25 (3H, s); ESIMS *m/z* 373.1 [M + H]$^+$ (calcd for $C_{20}H_{21}O_7$, 373.1).

4-Deoxy-12-hydroxynorherqueinone (**2a**): $[\alpha]_D^{25}$ −20 (*c* 0.3, CHCl$_3$); ^{1}H NMR (DMSO-*d$_6$*, 400 MHz) δ_H 14.32 (1H, s), 13.52 (1H, s), 9.28 (1H, s), 8.72 (1H, s), 4.64 (1H, q, *J* = 6.5 Hz), 2.66 (3H, s), 1.49 (3H, s), 1.38 (3H, d, *J* = 6.5 Hz), 1.26 (3H, s); ESIMS *m/z* 359.1 [M + H]$^+$ (calcd for $C_{19}H_{19}O_7$, 359.1).

4-Deoxy-*ent*-isoherqueinone (**3a**): $[\alpha]_D^{25}$ +34 (*c* 0.5, CHCl$_3$); ^{1}H NMR (DMSO-*d$_6$*, 400 MHz) δ_H 13.52 (1H, s), 8.13 (1H, s), 7.48 (1H, br s), 7.14 (1H, br s), 4.69 (1H, q, *J* = 6.5 Hz), 3.13 (3H, s), 2.66 (3H, s), 1.47 (3H, s), 1.42 (3H, d, *J* = 6.5 Hz), 1.22 (3H, s); ESIMS *m/z* 357.3 [M + H]$^+$ (calcd for $C_{20}H_{21}O_6$, 357.3).

4-Deoxy-oxopropylisoherqueinone A (**4a**): $[\alpha]_D^{25}$ +8 (*c* 0.5, CHCl$_3$), +10 (*c* 0.5, MeOH); ^{1}H NMR (DMSO-*d$_6$*, 400 MHz) δ_H 13.27 (1H, s), 8.48 (1H, s), 6.18 (1H, s), 5.73 (1H, s), 4.13 (1H, q, *J* = 6.5 Hz), 3.16 (2H, s), 2.80 (3H, s), 2.05 (3H, s), 1.35 (3H, s), 1.25 (3H, d, *J* = 6.5 Hz), 1.05 (3H, s); ESIMS *m/z* 399.1 [M + H]$^+$ (calcd for $C_{22}H_{23}O_7$, 399.1).

4-Deoxy-oxopropylisoherqueinone B (**5a**): $[\alpha]_D^{25}$ +3 (*c* 0.5, CHCl$_3$), +5 (*c* 0.5, MeOH); ^{1}H NMR (DMSO-*d$_6$*, 400 MHz) δ_H 13.27 (1H, s), 8.47 (1H, s), 6.18 (1H, s), 5.71 (1H, s) 4.10 (1H, q, *J* = 6.5 Hz), 3.16 (2H, s), 2.80 (3H, s), 2.05 (3H, s), 1.34 (3H, s), 1.26 (3H, d, *J* = 6.5 Hz), 1.05 (3H, s); ESIMS *m/z* 399.1 [M + H]$^+$ (calcd for $C_{22}H_{23}O_7$, 399.1).

Sclerodin (**6a = 10**): $[\alpha]_D^{25}$ −18 (*c* 0.5, CHCl$_3$); ^{1}H NMR (CDCl$_3$, 400 MHz) δ_H 11.5 (1H, s), 6.75 (1H, s), 5.06 (1H, q, *J* = 6.5 Hz), 2.69 (3H, s), 1.48 (3H, d, *J* = 6.5 Hz), 1.41 (3H, s), 0.92 (3H, s); ESIMS *m/z* 329.1 [M + H]$^+$ (calcd for $C_{18}H_{17}O_6$, 329.1).

3.5. Acetylation of 4-Deoxy-ent-peniciherqueinone (**1a**)

To a solution of 3.0 mg (2.7 mM) of **1a** in 3.0 mL of pyridine was added 0.4 mL of Ac$_2$O. After stirring the mixture for 4 h at room temperature, the pyridine and excess Ac$_2$O were removed under vacuum. Purification by analytical HPLC (YMC-ODS column, 4.6 × 250 mm; 0.7 mL/min; H$_2$O-MeCN (40:60)) yielded 4-deoxy-9,11,12-triacetyl-*ent*-peniciherqueinone (**1b**) (t_R = 35.8 min): $[\alpha]_D^{25}$ −35 (*c* 0.5,

CHCl$_3$), −42 (*c* 0.5, MeOH); ^{1}H NMR (CDCl$_3$, 400 MHz) δ_H 4.72 (1H, q, *J* = 6.5 Hz), 4.05 (3H, s), 2.71 (3H, s), 2.404 (3H, s), 2.401 (3H, s), 2.39 (3H, s), 1.59 (3H, s), 1.50 (3H, d, *J* = 6.5 Hz), 1.35 (3H, s); ESIMS *m/z* 499.5 [M + H]$^+$ (calcd for C$_{26}$H$_{27}$O$_{10}$, 499.5).

*3.6. Dehydration of Herqueinones (**4a**, **5a**, and **7**)*

To a solution of 0.5 mg (44 mM) of Na in 500 μL of anhydrous ethanol was added 1.5 mg (7.5 mM) of **4a** under nitrogen atmosphere. After stirring the mixture for 6 h at room temperature, the solvent was removed under vacuum. Purification by analytical HPLC (YMC-ODS column, 4.6 × 250 mm; 0.7 mL/min; H$_2$O-MeCN (40:60)) afforded the 8(6′)-dehydroxy derivative (**4b**) (t_R = 12.2 min) as a pure compound. Compounds **5a** and **7** were dehydrated to **5b** and **7a**, respectively, in the same manner.

4-Deoxy-8(6′)-dehydroxyoxopropylisoherqueinone A (**4b**): [α]$_D^{25}$ −27 (*c* 0.5, CHCl$_3$); ^{1}H NMR (CDCl$_3$, 400 MHz) δ_H 13.31 (1H, s), 12.79 (1H, s), 6.33 (1H, s), 5.61 (1H, s) 4.42 (1H, q, *J* = 6.5 Hz), 2.56 (3H, s), 2.38 (3H, s), 1.46 (3H, s), 1.38 (3H, d, *J* = 6.5 Hz), 1.20 (3H, s); ESIMS *m/z* 381.1 [M + H]$^+$ (calcd for C$_{22}$H$_{21}$O$_6$, 381.1).

4-Deoxy-8(6′)-dehydroxyoxopropylisoherqueinone B (**5b**): [α]$_D^{25}$ −26 (*c* 0.5, CHCl$_3$); ^{1}H NMR (CDCl$_3$, 400 MHz) δ_H 13.31 (1H, s), 12.79 (1H, s), 6.33 (1H, s), 5.61 (1H, s) 4.42 (1H, q, *J* = 6.5 Hz), 2.56 (3H, s), 2.38 (3H, s), 1.46 (3H, s), 1.38 (3H, d, *J* = 6.5 Hz), 1.20 (3H, s); ESIMS *m/z* 381.1 [M + H]$^+$ (calcd for C$_{22}$H$_{21}$O$_6$, 381.1).

8(6′)-Dehydroxy derivative of **7** (**7a**): [α]$_D^{25}$ −26 (*c* 0.5, CHCl$_3$); ^{1}H NMR (CDCl$_3$, 400 MHz) δ_H 13.31 (1H, s), 12.79 (1H, s), 6.33 (1H, s), 5.61 (1H, s) 4.42 (1H, q, *J* = 6.5 Hz), 2.56 (3H, s), 2.38 (3H, s), 1.46 (3H, s), 1.38 (3H, d, *J* = 6.5 Hz), 1.20 (3H, s); ESIMS *m/z* 381.1 [M + H]$^+$ (calcd for C$_{22}$H$_{21}$O$_6$, 381.1).

3.7. ECD Calcualtions

The conformational searches for the C-8 position of **4** and **5** were performed using Macromodel (Version 9.9, Schrodinger LLC.) software with "Mixed torsional/Low Mode sampling" in the GAFF force field. The experiments were conducted in the gas phase with a 50 kJ/mol energy window limit and a maximum of 10,000 steps to thoroughly examine all low-energy conformers. The Polak–Ribiere conjugate gradient (PRCG) method was utilized for minimization processes with 10,000 maximum iterations and a 0.001 kJ (mol Å)$^{-1}$ convergence threshold on the RMS gradient. Conformers within 10 kJ/mol of each global minimum for *R* and *S* form of **4** and **5** were used for gauge-independent atomic orbital (GIAO) shielding constant calculations without geometry optimization employing TmoleX Version 4.2.1 (COSMOlogic GmbH & Co. KG, Leverkusen, Germany) at the B3LYP/6-31G(d,p) level in the gas phase. The ECD spectra were simulated by overlapping each transition, where σ is the width of the band at $1/e$ height. ΔE_i and R_i are the excitation energies and rotatory strengths, respectively, for transition *i*. In the current work, the value was 0.10 eV.

$$\Delta \in (E) = \frac{1}{2.297 \times 10^{-39}} \frac{1}{\sqrt{2\pi\sigma}} \sum_{i}^{A} \Delta E_i R_i e^{[-(E-\Delta E_i)^2/(2\sigma)^2]}$$

3.8. Cytotoxicity and Antibacterial Assays

The cytotoxicity assay was performed in accordance with the published protocols [24]. The antimicrobial assay was performed according to the method described previously [25].

3.8.1. iNOS Assay

Mouse macrophage RAW 264.7 cells obtained from the American Type Culture Collection (ATCC, Rockville, MD, USA) were cultured in Dulbecco's modified Eagle's medium (DMEM) supplemented with 10% heat-inactivated fetal bovine serum (FBS) with antibiotics-antimycotics (PSF; 100 units/mL penicillin G sodium, 100 ng/mL streptomycin, and 250 ng/mL amphotericin B) [26,27]. The cells were seeded in 24-well plates (2×10^5 cells/mL). The next day, the culture media was changed to 1% FBS-DMEM, and the samples were treated with the test compounds. After pretreatment with the drug for 1 h, 1 μg/mL lipopolysaccharides (LPS) was added to stimulate NO production. The cells were incubated for an additional 18 h, and the amount of NO produced in the supernatant was determined by Griess reaction. Then, the absorbance was measured at 540 nm, and the nitrite concentration was determined by comparison with a sodium nitrite standard curve. The percent inhibition was calculated using the following formula: [1 − (NO level of test samples/NO levels of vehicle-treated control)] × 100. The IC_{50} values were calculated through nonlinear regression analysis using TableCurve 2-D v5.01 (Systat Software Inc., San Jose, CA, USA). At the same time, MTT assays were also performed to test cell viability. MTT solution (final concentration of 500 μg/mL) was added to the cells, and they were incubated for 4 h at 37 °C. The culture media was removed, and the remaining dyes were dissolved in DMSO. The absorbance of each well was measured at 570 nm using a VersaMax ELISA microplate reader (Molecular Devices, Sunnyvale, CA, USA). The percent survival was determined by comparison with a control group (LPS+).

3.8.2. Tube Formation Assay

Human umbilical vascular endothelial cells (HUVECs) were purchased from the American Type Culture Collection (ATCC, Rockville, MD, USA), and cultured in EGM-2 (Lonza, Walkersville, MD, USA) supplemented with 10% FBS and antibiotics-antimycotics (PSF) [28,29]. The cells were maintained at 37 °C under a humidified atmosphere containing 5% CO_2. A 96-well plate was coated with Matrigel (Corning) for 30 min at 37 °C under a humidified atmosphere containing 5% CO_2. HUVECs (1.8×10^4 cells/well) were mixed with the test compounds in 0.5% FBS EBM-2 medium with VEGF (50 ng/mL) or 0.5% FBS EBM-2 medium only for the VEGF negative control. The cells were incubated for 6 h and photographed using an inverted microscope (Olympus Optical Co. Ltd., Tokyo, Japan). Images were quantified with an angiogenesis analyzer using ImageJ software. Tube formation activity was calculated using the following formula: [(Total segment # (tested compound) − Total segment # (VEGF−)]/[Total segment # (VEGF+) − Total segment # (VEGF−)] × 100 (# stands for tubule segment number). The IC_{50} values were calculated through nonlinear regression analysis using TableCurve 2-D v5.01 (Systat Software Inc., San Jose, CA, USA). Cell viabilities were evaluated with the MTT assay. HUVECs (0.8×10^4 cells/well) were seeded into a 96-well plate and indicated for 1 day. The culture medium was replaced with serum-free medium, and the cells were incubated overnight. After starvation, the cells were treated with the test compounds and VEGF (50 ng/mL) in 2% FBS EBM-2 medium. Cells were incubated for a further 24 h, and MTT solution (final concentration of 500 μg/mL) was added to the cells to measure the cell viability. The formazan products were dissolved in DMSO. The absorbance of each well was measured at 570 nm using a VersaMax ELISA microplate reader (Molecular Devices, Sunnyvale, CA, USA).

3.8.3. Adiponectin Production Assay

Human bone marrow-mesenchymal stem cells (hBM-MSCs) were purchased from Lonza, Inc. (Walkersville, MD, USA) and cultured in low-glucose (1 g/L) DMEM supplemented with 10% FBS, penicillin-streptomycin, and Glutamax^{TM} (Invitrogen, Carlsbad, CA, USA). To induce adipogenesis, the cell growth medium was replaced with high-glucose (4.5 g/L) DMEM supplemented with 10% FBS, penicillin-streptomycin, 10 μg/mL insulin, 0.5 μM dexamethasone, and 0.5 mM 3-isobutyl-1-methylxanthine (IBMX) (IDX conditions) [30]. IBMX, pioglitazone, and aspirin were purchased from Sigma-Aldrich (St. Louis, MO, USA). hBM-MSCs were stained with 0.2% oil red O (ORO) reagent for 10 min at 24 °C, and then washed with H_2O four times. Following a 10-min elution with isopropanol, the absorbance was measured at 500 nm using a spectrophotometer. To visualize the nucleus, the hBM-MSCs were counterstained with hematoxylin reagent for 2 min and then washed twice with H_2O. The level of adipocyte differentiation was observed and counted using an inverted phase microscope. A Quantikine immunoassay kit (R&D Systems, Minneapolis, MN, USA) was used for quantitative determination of adiponectin in the cell culture supernatants.

3.8.4. Receptor Binding Assay

The time-resolved fluorescence resonance energy transfer (TR-FRET)-based nuclear receptor binding assay to evaluate binding of the ligand to PPARγ was performed using Lanthanscreen™ competitive binding assay kits (Invitrogen) [30]. All assay measurements were performed using a CLARIOstar instrument (BMG LABTECH, Ortenberg, Germany) with the settings described in the TR-FRET manufacturer's instructions.

4. Conclusions

Six new phenalenone derivatives (**1–6**) and five known compounds (**7–11**) of the herqueinone class were isolated from a marine-derived fungus *Penicillium* sp. The structure elucidation of compounds **1–6** were established by combined spectroscopic methods. The absolute configurations were determined by chemical modifications and their specific rotations. 4-Hydroxysclerodin (**6**) and an acetone adduct of a triketone (**7**) exhibited moderate anti-angiogenetic and anti-inflammatory activities, respectively, while *ent*-peniciherqueinone (**1**) and isoherqueinone (**9**) exhibited moderate abilities to induce adipogenesis without cytotoxicity.

5. Patents

Shin, J.; Oh, K.-B. Phenalenone derivatives and antimicrobial composition. KR 2014112273 A 20140923, 2014.

Supplementary Materials: The following are available online at http://www.mdpi.com/1660-3397/17/3/176/s1. Figures S1–S41: annotated 1D NMR and selected 2D NMR spectra of **1–6**, Figures S42–S47: ^{1}H NMR spectra of **1a–6a**, Figures S48–S50: ^{1}H NMR spectra of **4b**, **5b**, and **7a**, Figures S51 and S52: The CD and ECD spectrum of **4–5**, Figure S53: The time-scale LC-MS analysis of **7**, Figure S54: iNOS assay and MTT assay, Figure S55: Tube formation assay and MTT assay, Figure S56: adiponectin production assay and PPARγ binding assay, Table S1: The results of bioactivity tests.

Author Contributions: S.C.P. and E.J. carried out the isolation and structural elucidation; K.-B.O. performed the antimicrobial bioassays; S.A. and M.N. performed the adiponectin production and receptor binding assays; D.K. and S.K.L. performed iNOS and tube formation assays; J.S. and D.-C.O. reviewed and evaluated all data; J.S. and K.-B.O. supervised the research work and prepared the paper.

Funding: This study was supported by the National Research Foundation (NRF, grant No. 2018R1A4A1021703) funded by the Ministry of Science, ICT, & Future Planning, Korea.

Acknowledgments: We thank the Basic Science Research Institute in Daegu, Korea, for providing mass spectrometric data.

Conflicts of Interest: The authors declare no conflict of interest.

References

1. Elsebai, M.F.; Saleem, M.; Tejesvi, M.V.; Kajula, M.; Mattila, S.; Mehiri, M.; Turpeinen, A.; Pirttila, A.M. Fungal phenalenones: chemistry, biology, biosynthesis and phylogeny. *Nat. Prod. Rep.* **2014**, *31*, 628–645. [CrossRef] [PubMed]

2. Nazir, M.; Maddah, F.; Kehraus, S.; Egereva, E.; Piel, J.; Brachmann, A.O.; König, G.M. Phenalenones: Insight into the biosynthesis of polyketides from the marine alga-derived fungus *Coniothyrium* cereal. *Org. Biomol. Chem.* **2015**, *13*, 8071–8079. [CrossRef] [PubMed]

3. Galarraga, J.A.; Neill, K.G.; Raistrick, H. Colouring matters of *Penicillium herquei*. *Biochem. J.* **1955**, *61*, 456–464. [CrossRef] [PubMed]

4. Frost, D.A.; Halton, D.D.; Morrison, G.A. Naturally occurring compounds related to phenalenone. Part 8. Structure and synthesis of demethylherqueichrysin. *J. Chem. Soc. Perkin Trans.* **1977**, 2443–2448. [CrossRef]

5. Narasimhachari, N.; Vining, L.C. Herqueichrysin, a new phenalenone antibiotic from *Penicillium herquei*. *J. Antibiot.* **1972**, *25*, 155–162. [CrossRef]

6. Julianti, E.; Lee, J.-H.; Liao, L.; Park, W.; Park, S.; Oh, D.-C.; Oh, K.-B.; Shin, J. New Polyaromatic Metabolites from a marine-Derived Fungus *Penicillium* sp. *Org. Lett.* **2013**, *15*, 1286–1289. [CrossRef]

7. Furihata, K.; Seto, H. Decoupled HMBC (D-HMBC), an improved technique of HMBC. *Tetrahedron Lett.* **1995**, *36*, 2817–2820. [CrossRef]

8. Tansakul, C.; Rukachaisirikul, V.; Maha, A.; Kongprapan, T.; Phongpaichit, S.; Hutadilok-Towatana, N.; Borwornwiriyapan, K.; Sakayaroj, J. A new phenalenone derivative from the soil fungus *Penicillium herquei* PSU-RSPG93. *J. Nat. Prod. Res.* **2014**, *28*, 1718–1724. [CrossRef]

9. Barton, D.H.R.; Mayo, P.; Morrison, G.A.; Raistrick, H. The constitutions of atrovenetin and of some related herqueinone derivatives. *Tetrahedron* **1959**, *6*, 48–62. [CrossRef]

10. Brooks, J.S.; Morrison, G.A. The constitution of herqueinone and its relationship to isoherqueinone. *Tetrahedron Lett.* **1970**, *12*, 963–966. [CrossRef]

11. Rani, B.R.; Ubukata, M.; Osada, H. Reduction of arylcarbonyl using zinc dust in acetic acid. *Bull. Chem. Soc. Jpn.* **1995**, *68*, 282–284. [CrossRef]

12. Bonner, T.G.; McNamara, P. The pyridine-catalysed acetylation of phenols and alcohols by acetic anhydride. *J. Chem. Soc. B* **1968**, 795–797. [CrossRef]

13. Lugemwa, F.N.; Shaikh, K.; Hochstedt, E. Facile and efficient acetylation of primary alcohols and phenols with acetic anhydride catalyzed by dried Sodium bicarbonate. *Catalysts* **2013**, *3*, 954–965. [CrossRef]

14. Ayer, W.A.; Hoyano, Y.; Pedras, M.S.; Altena, I. Metabolites produced by the Scleroderris canker fungus, *Gremmeniellaabietina*. Part 1. *Can. J. Chem.* **1986**, *64*, 1585–1589. [CrossRef]

15. Krohn, K.; Sohrab, M.D.H.; Aust, H.-J.; Draeger, S.; Schulz, B. Biologically active metabolites from fungi, 19: New isocoumarins and highly substituted benzoic acids from the endophytic fungus, *Scytalidium* sp. *Nat. Prod. Res.* **2004**, *18*, 277–285. [CrossRef]

16. Homma, K.; Fukuyama, K.; Katsube, Y.; Kimura, Y.; Hamasaki, T. structure and absolute configuration of an atrovenetin-like metabolite from *Aspergillus silvaticus*. *Agric. Biol. Chem.* **1980**, *44*, 1333–1337. [CrossRef]

17. Stodola, F.H.; Raper, K.B.; Fennell, D.I. Pigments of *Penicillium herquei*. *Nature* **1951**, *167*, 773–774. [CrossRef]

18. Suga, T.; Yoshioka, T.; Hirata, T.; Aoki, T. ^{13}C NMR Signal Assignments of Herqueinone and Its Phenalenone Derivatives. *Bull. Chem. Soc. Jpn.* **1983**, *56*, 3661–3666. [CrossRef]

19. Harman, R.E.; Cason, J.; Stodola, F.H.; Adkins, A. Structural features of herqueinone, a red pigment from *Penicillium herquei*. *J. Org. Chem.* **1955**, *20*, 1260–1269. [CrossRef]

20. Cason, J.; Koch, C.W.; Correia, J.S. Structures of herqueinone, isoherqueinone and norherqueinone. *J. Org. Chem.* **1970**, *35*, 179–186. [CrossRef]

21. Ayer, W.A.; Hoyano, Y.; Pedras, M.S.; Clardy, J.; Arnold, E. Metabolites produced by the scleroderris canker fungus, *Gremmeniellaabietina*. Part 2. The structure of scleroderolide. *Can. J. Chem.* **1987**, *65*, 748–753. [CrossRef]

22. Ayer, W.A.; Kamada, M.; Ma, Y.T. Sclerodin and related compounds from a plant disease causing fungus. *Scleroderris yellow*. *Can. J. Chem.* **1989**, *67*, 2089–2094. [CrossRef]

23. Shiomi, K.; Matsui, R.; Isozaki, M.; Chiba, H.; Sugai, T.; Yamaguchi, Y.; Masuma, R.; Tomoda, H.; Chiba, T.; Yan, H.; et al. Fungal Phenalenones Inhibit HIV-1 Integrase. *J. Antibiot.* **2005**, *58*, 65–68. [CrossRef] [PubMed]

24. Vichai, V.; Kirtikara, K. Sulforhodamine B colorimetric assay for cytotoxicity screening. *Nat. Protoc.* **2006**, *1*, 1112–1116. [CrossRef]

25. Kim, C.-K.; Woo, J.-K.; Kim, S.-H.; Cho, E.; Lee, Y.-J.; Lee, H.-S.; Sim, C.J.; Oh, D.-C.; Oh, K.-B.; Shin, J. Meroterpenoids from a Tropical *Dysidea* sp. *Sponge*. *J. Nat. Prod.* **2015**, *78*, 2814–2821. [CrossRef] [PubMed]

26. Nakane, M.; Klinghofer, V.; Kuk, J.E.; Donnelly, J.L.; Budzik, G.P.; Pollock, J.S.; Basha, F.; Carter, G.W. Novel potent and selective inhibitors of inducible nitric oxide synthase. *Mol. Pharmacol.* **1995**, *47*, 831–834. [PubMed]

27. Chung, H.-J.; Koh, W.; Kim, W.K.; Shin, J.-S.; Lee, J.; Lee, S.K.; Ha, I.-H. The anti-inflammatory effects of Shinbaro3 is mediated by downregulation of the TLR4 signalling pathway in LPS-stimulated RAW 264.7 macrophages. *Mediat. Inflamm.* **2018**. [CrossRef]

28. Carmeliet, P.; Jain, R.K. Angiogenesis in cancer and other diseases. *Nature* **2000**, *407*, 249–257. [CrossRef]

29. Yu, S.; Oh, J.; Li, F.; Kwon, Y.; Cho, H.; Shin, J.; Lee, S.K.; Kim, S. New scaffold for angiogenesis inhibitors discovered by targeted chemical transformations of wondonin natural products. *ACS Med. Chem. Lett.* **2017**, *8*, 1066–1071. [CrossRef]
30. Anh, S.; Lee, M.; An, S.; Hyun, S.; Hwang, J.; Lee, J.; Noh, M. 2-Formyl-komarovicine promotes adiponectin production in human mesenchymal stem cells through PPARγ partial agonism. *Bioorg. Med. Chem.* **2018**, *26*, 1069–1075. [CrossRef]

 *marine drugs*

Article

New *Bis*-Alkenoic Acid Derivatives from a Marine-Derived Fungus *Fusarium solani* H915

Shun-Zhi Liu [1,†], Xia Yan [2,†], Xi-Xiang Tang [3], Jin-Guo Lin [1,*] and Ying-Kun Qiu [4,*]

[1] College of Material Engineering, Fujian Agriculture and Forestry University, Fuzhou, Fujian 350002, China; shunzhi0306@126.com

[2] Li Dak Sum Yip Yio Chin Kenneth Li Marine Biopharmaceutical Research Center, Ningbo University, Ningbo 315832, China; yanxia@nbu.edu.cn

[3] Key Laboratory of Marine Biogenetic Resources, Third Institute of Oceanography State Oceanic Administration, Xiamen 361005, China; tangxixiang@tio.org.cn

[4] Fujian Provincial Key Laboratory of Innovative Drug Target Research, School of Pharmaceutical Sciences, Xiamen University, South Xiang-An Road, Xiamen 361102, China

* Correspondence: fjlinjg@126.com (J.-G.L.); qyk@xmu.edu.cn (Y.-K.Q.); Tel./Fax: +86-592-2189868 (Y.-K.Q.)

† These authors contribute equally to this paper.

Received: 20 September 2018; Accepted: 30 November 2018; Published: 3 December 2018

Abstract: *Fusarium solani* H915 is a fungus derived from mangrove sediments. From its ethyl acetate extract, a new alkenoic acid, fusaridioic acid A (**1**), three new *bis*-alkenoic acid esters, namely, fusariumester A_1 (**2**), A_2 (**3**) and B (**4**), together with three known compounds (**5–7**), were isolated. The structures of the new compounds were comprehensively characterized by high resolution electrospray ionization-mass spectrometry (HR-ESI-MS), 1D and 2D nuclear magnetic resonance (NMR). Additionally, the antifungal activities against tea pathogenic fungi *Pestalotiopsis theae* and *Colletotrichum gloeosporioides* were studied. The new compound, **4**, containing a β-lactone ring, exhibited moderate inhibitory activity against *P. theae*, with an MIC of 50 µg/disc. Hymeglusin (**6**), a typical β-lactone antibiotic and a terpenoid alkaloid, equisetin (**7**), exhibited potent inhibitory activities against both fungal species. The isolated compounds were evaluated for their effects on zebrafish embryo development. Equisetin clearly imparted toxic effect on zebrafish even at low concentrations. However, none of the alkenoic acid derivatives exhibited significant toxicity to zebrafish eggs, embryos, or larvae. Thus, the β-lactone containing alkenoic acid derivatives from *F. solani* H915 are low in toxicity and are potent antifungal agents against tea pathogenic fungi.

Keywords: *Fusarium solani* H915; *bis*-alkenoic acid esters; fusaridioic acid A; fusariumester A_1; fusariumester A_2; fusariumester B; tea pathogenic fungi inhibitory effect

1. Introduction

Mangrove areas commonly experience frequent tides, strong winds, high water temperatures and exposure to strong ultraviolet radiation, while mangrove sediments typically contain high mineral composition, can readily become polluted and may be characterized as acidic, hypoxic and/or oligotrophic environments [1]. Because of the unique habitat, the discovery of new lead compounds with antibacterial activity from mangrove microorganisms has rapidly become a hot topic in the field of natural products research [2].

Tea (*Camellia sinensis* O. Kuntze) is an important economic crop in many Asian and African countries. However, tea production has been hindered by various biotic and abiotic factors. For instance, fungal diseases, especially those infecting the leaves, are among the biotic factors that cause severe damage, thereby resulting in high yield losses [3,4]. The major fungal diseases of tea leaves include blister blight (*Exobasidium vexans Massee*), grey blight (*Pestalotiopsis theae* (Sawada) Steyaert), brown blight (*Colletotrichum camelliae* Massee), sooty mold (*Capnodium theae* Boedijn) and red rust (*Cephaleuros parasiticus* Karst) [3,5–9].

To date, reports on bioactive compounds against tea pathogenic fungi are limited. Previously, we identified two new macrolactins [10,11] from *Bacillus subtilis* B5, a bacterium isolated from the 3000 m deep sea sediment of the Pacific Ocean that exhibit antifungal activity against *P. theae* and *C. gloeosporioides*. In the present study, *Fusarium solani* H915, a fungal strain that originated from the mangrove sediment of the Zhangjiangkou Mangrove National Nature Reserve, was found to possess antifungal activity. In this study, a new alkenoic acid, fusaridioic acid A (**1**), together with three new *bis*-alkenoic acid esters (Figure 1), namely, fusariumester A_1 (**2**), fusariumester A_2 (**3**) and fusariumester B (**4**) and three known metabolites, namely, L660282 (**5**) [12], hymeglusin (**6**) [13] and equisetin (**7**) [14], were isolated from the ethyl acetate extract of a culture of *F. solani* H915. The structures of the new compounds were comprehensively characterized by HR-ESI-MS, ^{1}H NMR, ^{13}C NMR and 2D-NMR. Their antifungal activities against tea pathogenic fungi *P. theae* and *C. gloeosporioides* were studied. Equisetin and the alkenoic acid derivatives containing a β-lactone ring exhibited inhibitory activities against the fungi. However, equisetin exhibited strong anti-proliferative effects on zebrafish embryos and larvae. The alkenoic acid derivatives with β-lactone rings from *F. solani* H915 are low in toxicity and thus may be potentially used as antifungal agents against tea pathogenic fungi.

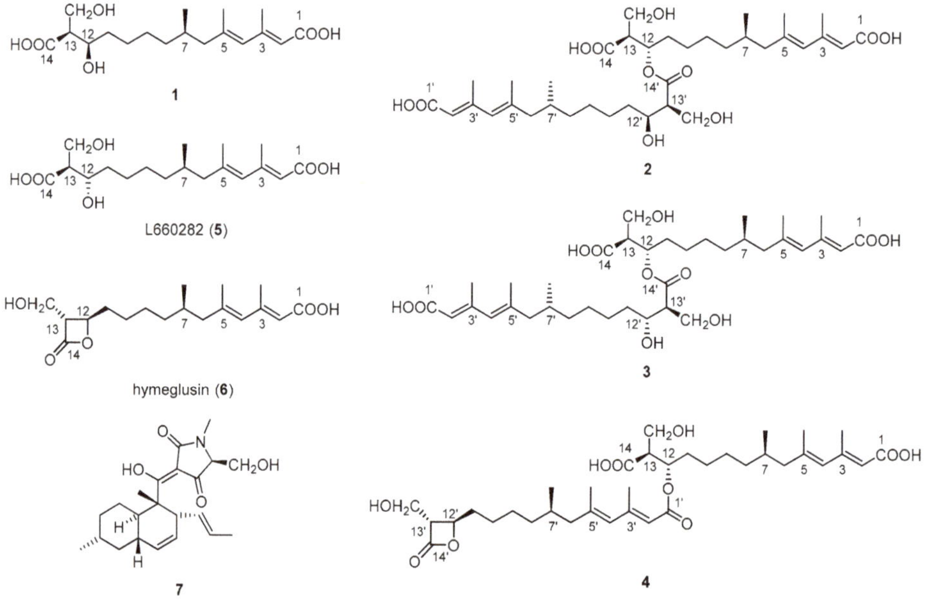

Figure 1. Structures of compounds 1–7 isolated from an extract of *Fusarium solani* H915.

2. Results

2.1. Structural Identification of New Compounds

Fusaridioic acid A (**1**) was isolated as a white amorphous powder. The infrared (IR) spectrum of **1** indicated the presence of free and conjugated carboxylic acid carbonyl signals at 1734 cm^{-1} and 1683 cm^{-1}, respectively. The UV maximum absorption wavelengths at λ_{max} (log ε), 233 (3.58) nm and 266 (3.80) nm, belong to an unconjugated carbonyl and another conjugated carbonyl, respectively. Its molecular formula of $C_{18}H_{30}O_6$, which gave four unsaturation degrees, was established by the HR-ESI-MS at 341.1968 $[M - H^+]^-$ (calcd. for 341.1964 $C_{18}H_{29}O_6$) and by the data of 1D-NMR. In the low field of ^{1}H NMR, two olefinic proton signals were observed with br. s peaks at δ_H 5.50 (H-2) and 5.67 (H-4). In the sp^3 region of the ^{1}H NMR spectrum, two methyl groups at δ_H 2.09 (d, J = 1.1 Hz, 3-CH$_3$) and 1.70 (d, J = 1.1 Hz, 5-CH$_3$) were considered linked to quaternary olefinic carbons. Another methyl at δ_H 0.73 (d, J = 6.6 Hz, 7-CH$_3$) is connected to a methylene. In the sp^2 region of ^{13}C NMR a carboxyl signal emerged at δ_C 174.9 (C-14), whereas another one (which is conjugated with the double-bond system) presented at δ_C 168.1 (C-1). The sp^3 high-field region of the ^{1}H NMR spectrum showed the existence of three methyl proton signals. Two methyl groups with J values of 1.1 Hz at δ_H 2.09 (3-CH$_3$) and 1.70 (5-CH$_3$) were considered linked to quaternary olefinic carbons. The other methyl at δ_H 0.73 (d, J = 6.6 Hz, 7-CH$_3$) is connected to a methylene. In addition, a methine and a methylene, bound to oxygen atoms, could be found in the ^{13}C NMR and distortionless enhancement by polarization transfer (DEPT) spectra at δ_C 69.3 (C-12) and 55.4 (13-CH$_2$OH). With the aid of ^{1}H-^{1}H homonuclear chemical shift correlation spectroscopy (COSY) spectra, the proton signal at δ_H 3.55 (m) was attributed to H-12. The two dd peaks at δ_H 3.51 (J = 10.5, 8.4 Hz) and 3.46 (J = 10.4, 5.4 Hz), which form a typical ABX coupling system with H-13, were assigned to the two protons at 13-CH$_2$OH. Elucidation of heteronuclear single quantum coherence (HSQC), ^{1}H-^{1}H COSY and ^{1}H detected heteronuclear multiple bond correlation (HMBC) spectra, led to the planar structure (Figure 1), which is almost identical to that of L-660282 (**5**), a compound isolated from a culture of *Cephalosporium* sp. [12]. Indeed, most of the 1D NMR spectral data of **1** approached those of L-660282 (**5**) (Tables 1 and 2). The ^{13}C NMR signal differences between **1** and **5** are found at positions 11, 12, 13, 14 and 13-CH$_2$OH (δ_C 35.0, 69.3, 55.4, 174.9 and 60.1 in **1**; and δ_C 35.7, 69.5, 56.0, 175.2 and 61.3 in **5**), revealing that **1** is an epimer of **5** at 12-OH. In both compounds, the C-14 carbonyl oxygen forms a hydrogen bond to 12-OH. At the same time, the 14-carboxyl hydroxyl group can form a hydrogen bond with the 13-CH$_2$OH group. Both those H-bonding interactions will lead to six-membered rings for both isomers. In compound **5**, H-12 and H-13 are in a trans-coplanar position, inducing an H12-H13 coupling constant of 8.3 Hz (due to the presence of such electron withdrawing groups as hydroxyl and carbonyl, the coupling constant is slightly lower than the Karplus equation prediction coupling value). However, the bond angle between H-12 and H-13 in **1** is about 60°, leading to a reduction of the $J_{\text{H-12, 13}}$ value to 5.7 Hz (Figure 2). Thus, the coupling splittings of H-13 in **5** present a td peak with J values of 8.5 Hz $\times$ 2 and 4.6 Hz $\times$ 1, whereas those in **1** led to a dt peak with J values of 8.3 Hz $\times$ 1 and 5.7 Hz $\times$ 1. The structural connection of **1** was further confirmed by the HSQC, ^{1}H-^{1}H COSY and HMBC spectra (Figure 3). The configuration of the two double bonds was revealed by the nuclear overhauser effect spectroscopy (NOESY) spectrum (Supplementary Materials Figures S1–S8).

The absolute configuration of C-7, C-12 and C-13 in **1** was revealed by alkaline hydrolysis of hymeglusin (**6**), a typical β-lactone antibiotic previously isolated from a culture of *Scopulariopsis* sp. F-244 [13]. Alkaline hydrolysis of hymeglusin (**6**), which mainly experienced a bi-molecular substitution nucleophilic (SN2) reaction process, yielded a single product, whose thin layer chromatography (TLC) retardation factor (Rf) value, high performance liquid chromatography (HPLC) retention time and optical rotation ($[\alpha]_D^{25}$) value were close to those of compound **1**, indicating that the absolute configuration of **1** should be 7*R*, 12*R*, 13*S*.

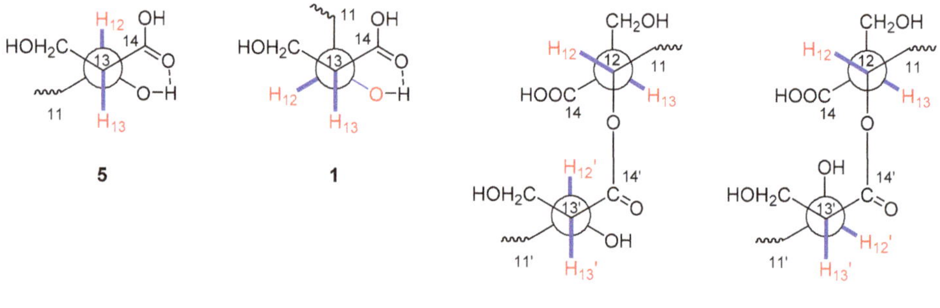

Figure 2. Key preferential conformations of **1–3** and **5**.

Fusariumester A$_1$ (**2**) was isolated as colorless viscous oil, whose $[\alpha]_D^{25}$ value was close to 0° (*c* = 0.1, CH$_3$OH). Its molecular formula of C$_{36}$H$_{58}$O$_{11}$, which gave eight unsaturation degrees, was established by the HRESIMS at 665.3883 [M − H$^+$]$^-$ (calcd. for 665.3901 C$_{36}$H$_{57}$O$_{11}$) and by the data of 1D NMR. The IR spectrum showed a wide and strong adsorption signal band at 1695 cm^{-1}, caused by overlapping of multiple carbonyl signals. Absorption at 1631 cm^{-1} belongs to the conjugated C=C system. The ^{1}H and ^{13}C NMR data of **2**, whose signals emerged in duplicate pairs, indicated the presence of two similar structural units. The NMR data of each unit were close to those of **5**, indicating that **2** could be a dimer of **5**. In detail, four olefinic proton signals were found in the low field region of the ^{1}H NMR spectrum of **2**, at δ_H 5.70 (1H, br.s, H-4), 5.71 (1H, br.s, H-4′), 5.60 (1H, br.s, H-2) and 5.56 (1H, br.s, H-2′). With the aid of ^{13}C NMR, DEPT, HSQC and HMBC spectra, their corresponding carbon signals belonging to two pairs of conjugated dienes could be assigned. Three pairs of methyl signals were also found in the high-field region of the ^{1}H and ^{13}C NMR spectra. All these pairs of signals, together with a pair of carboxyls [δ_C 168.3 (C-1) and 168.2 (C-1′)] and some other sp^3 carbons (C-6, 7, 8, 9 and C-6′, 7′, 8′, 9′), were paired very well, with almost identical chemical shifts. However, the chemical shifts of the ^{1}H and ^{13}C atoms at positions 11-14 and 11′-14′ were different, indicating that the two structural units of **5** are linked with each other at these positions. In the ^{13}C NMR spectrum, δ_C 174.4 (C-14) differed from 173.3 (C-14′); δ_C 53.3 (C-13) differed from 56.2 (C-13′); δ_C 71.6 (C-12) differed from 69.3 (C-12′); δ_C 32.6 (C-11) differed from 35.8 (C-11′) and δ_C 60.6 (13-CH$_2$OH) differed from 61.5 (13′-CH$_2$OH). Compared with those of compound **5**, the low-field shift of C-12 and high-field shifts of C-13, C-11 and C-14′ revealed esterification between the 12-hydroxyl and 14′-carboxyl. This finding was further confirmed by the HMBC correlation between H-12 [δ_H 5.00 (1H, m)] and C-14′ (δ_C 173.3), as well as the low-field shifting of H-12. The relative position between H-12 and 13-CH$_2$OH and that between H-12′ and 13′-CH$_2$OH were both deduced as trans coplanar positions, by the coupling constants and the td peak splitting of H-13 and H-13′ in ^{1}H NMR (Figure 2). Alkaline hydrolysis of **2** yielded a single product, compound **5**, indicating that compound **2** is an esterified product of two molecules of compound **5**. The full assignment of ^{1}H (Table 1) and ^{13}C NMR data (Table 2) was deduced by careful elucidation of HSQC, ^{1}H-^{1}H COSY, HMBC and NOESY spectra (Figure 3, Supplementary Materials Figures S9–S17). The absolute configurations of C-7, 7′, C-12, 12′ and C-13, 13′ in **2** were identical to those in compound **5**, because the TLC Rf value, HPLC retention time and optical rotation value of the alkaline hydrolysis of **2** were almost identical to those of **5** (Supplementary Materials Figure S35). Considering that compounds **1** and **5** are a pair of epimers that differ at C-12 and their similar biosynthetic pathway, the absolute configuration of **2** should be 7*R*, 12*S*, 13*S* and 7′*R*, 12′*S*, 13′*S*.

Fusariumester A$_2$ (**3**) was isolated as colorless viscous oil. The ^{1}H and ^{13}C NMR data of **3** were almost identical with those of compound **2** except for the protons and carbons around C-13′. In detail, compared with those of **2**, chemical shift of H-12′ in **3** is low-field shifted from δ_H 3.50 to δ_H 3.62; that of H-13′ is low-field shifted from δ_H 2.44 to δ_H 2.50; and that of 13′-CH$_2$OH is high-field shifted from δ_H 3.76 and 3.61 to δ_H 3.61 and 3.56. In the ^{13}C NMR of **3**, the signals of C-13′, 14′ and 13′-CH$_2$OH are high-field shifted about 0.5–1.5 ppm. The structure of **3** was deduced to be an epimer of compound **2** at C-12′ and C-13′. This deduction is confirmed by the coupling constants and the dt peak splitting of H-13′ in ^{1}H NMR (Figure 2). Alkaline hydrolysis of **3** yielded a pair of products, compounds **1** and **5**, indicating that compound **3** is an esterified product of compound **1** and **5**. The absolute configuration of **3** is 7*R*, 12*R*, 13*S* and 7′*R*, 12′*S*, 13′*S*, which was confirmed by comparing the TLC Rf value, HPLC retention time and optical rotation value of the alkaline hydrolysis of **3** with those of compounds **1** and **5** (Supplementary Materials Figure S35).

Fusariumester B (**4**) is a colorless viscous oil. Its molecular formula of C$_{36}$H$_{56}$O$_{10}$ established by the HRESIMS quasi-molecular ion peak at *m/z* 647.3814 [M − H$^+$]$^-$ (calcd. for 647.3873 C$_{36}$H$_{55}$O$_{10}$), yields nine unsaturation degrees. The molecular weight of **4** is lower than those of **2** and **3** by 18 Da, indicating that **4** could be the closed-ring product of **2** or **3**. The IR carbonyl absorption at 1822 cm^{-1} indicated the presence of a β-lactone group. Carbonyl signals at 1715 and 1698 cm^{-1} were attributed to other carbonyl groups. Most of the ^{1}H, ^{13}C NMR and DEPT data of **4** were similar with those of **2** except the carbon signals assigned to C-1′, 12′, 13′, 14′ and 13′-CH$_2$OH. Indeed, the NMR data of subunit C-1′–C-14′ are close to those of hymeglusin (**6**), indicating the presence of β-lactone in **4**. The β-lactone structure leads to the chemical shift changing of the atoms at positions 12′, 13′, 14″ and 13′-CH$_2$OH. On the other hand, compared with the ^{13}C NMR data of hymeglusin reported in Reference [, the chemical shift of C-1′ was high-field shifted from 172.0 to 166.1 and that of C-2′ was low-field shifted from 116.7 to 118.6, indicating esterification of the 1′-carboxyl. Based on the correlation between H-12 [δ_H 5.05 (1H, q-like, *J* = 6.5 Hz)] and C-1′ (δ_C 166.1) in HMBC spectrum, 1′-carboxyl is confirmed to be esterified with 12-hydroxyl. Alkaline hydrolysis of **4** also yielded a pair of products, namely, compounds **1** and **5**. The open-ring diacid structure unit led to the yield of compound **1**, while the β-lactone structure unit underwent an open-ring SN2 alkaline hydrolysis ring opening to yield compound **5** as one of the final products (Supplementary Materials Figure S35). Base on the above findings, the absolute configuration of **4** is 7*R*, 12*S*, 13*S* and 7′*R*, 12′*R*, 13′*R*.

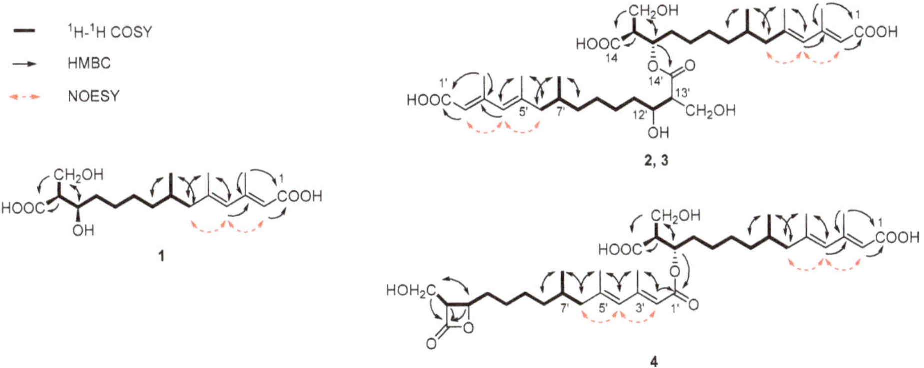

Figure 3. Key ^{1}H–^{1}H COSY, HMBC and NOESY correlations of **1**–**4**.

Table 1. ^{1}H nuclear magnetic resonance (NMR) (DMSO-d_6, 600 MHz) data of compounds **1–6**.

No.	1	2	3	4	5	6 *
2	5.50, br.s	5.60, br.s	5.58, br.s	5.56, br.s	5.50, s	
2′		5.56, br.s	5.57, br.s	5.59, br.s		5.57, br.s
4	5.67, br.s	5.70, br.s	5.71, br.s	5.71, br.s	5.66, s	
4′		5.71, br.s	5.72, br.s	5.76, br.s		5.73, br.s
6a	2.00, dd (13.0, 6.1)	2.02, m	2.05, dd (5.0, 3.3)	2.04, dd (13.1, 6.0)	2.00, dd (13.1, 6.0)	
6b	1.75, dd (13.3, 8.3)	1.82, m	1.82, m	1.80, dd (12.1, 4.4)	1.75, dd (13.1, 8.3)	
6′a		2.06, dd (13.6, 6.2)	2.08, m	2.08, dd (13.3, 6.0)		2.07, dd (13.2, 6.1)
6′b		1.79, m	1.79, m	1.83, dd (13.2, 8.4)		1.83, dd (12.8, 8.3)
7	1.58, m	1.60, m	1.62, m	1.61, m	1.57, br.dd (12.7, 6.4)	
7′		1.60, m	1.62, m	1.64, m		1.64, m
8a	1.18, m	1.24, m	1.24, m	1.25, m	1.19, m	
8b	1.01, m	1.06, m	1.05, m	1.05, m	1.00, m	
8′a		1.24, m	1.24, m	1.28, m		1.28, m
8′b		1.02, m	1.05, m	1.10, m		1.10, m
9a	1.20, m	1.19, m	1.25, m	1.27, m	1.26, m	
9b	1.20, m	1.19, m	1.25, m	1.20, m	1.12, m	
9′a		1.30, m	1.25, m	1.27, m		1.27, m
9′b		1.30, m	1.25, m	1.20, m		1.20, m
10a	1.29, m	1.32, m	1.35, m	1.36, m	1.36, m	
10b	1.20, m	1.20, m	1.19, m	1.20, m	1.15, m	
10′a		1.41, m	1.36, m	1.36, m		1.36, m
10′b		1.26, m	1.24, m			
11a	1.34, m	1.54, m	1.51, m	1.57, m	1.30, m	
11b	1.24, m			1.55, m	1.23, m	
11′a		1.31, m	1.37, m	1.80, m		1.80, m
11′b			1.31, m	1.74, m		1.73, m
12	3.55, m	5.00, m	5.00, td (7.8, 4.7)	5.05, q (6.5)	3.44, td (8.3, 2.9)	
12′		3.50, m	3.62, m	4.53, td (6.6, 4.2)		4.53, td (6.7, 4.2)
13	2.34, dt (8.3, 5.7)	2.57, td (8.3, 4.6)	2.60, td (8.5, 4.9)	2.65, td (8.0, 4.9)	2.29, td (8.5, 4.6)	
13′		2.44, td, (8.8, 4.6)	2.50, m	3.50, br.dd (7.8, 3.9)		3.50, br.dd (7.7, 3.9)
3-CH$_3$	2.09, d (1.1)	2.15, br.s	2.15, br.s	2.14, d (0.9)	2.09, d (1.1)	
3′-CH$_3$		2.14, br.s	2.16, br.s	2.18, br.s		2.16, d (1.1)
5-CH$_3$	1.70, d (1.1)	1.74, br.s	1.75, br.s	1.73, br.s	1.70, d (1.1)	
5′-CH$_3$		1.75, br.s	1.76, br.s	1.77, br.s		1.76, d (1.1)
7-CH$_3$	0.73, d (6.6)	0.78, d (6.6)	0.78, d (6.6)	0.77, d (6.6)	0.73, d (6.6)	
7′-CH$_3$		0.79, d (6.4)	0.79, d (6.6)	0.80, d (6.6)		0.80, d (6.6)
13-CH$_2$OH	3.51, dd (10.5, 8.4)	3.60, dd (10.4, 4.5)	3.58, dd (10.5, 9.0)	3.60, dd (10.1, 1.5)	3.66, dd (10.4, 4.5)	
	3.46, dd (10.4, 5.4)	3.53, dd (10.4, 9.0)	3.54, dd (10.5, 4.3)	3.52, dd (10.5, 4.8)	3.55, dd (10.2, 9.1)	
13′-CH$_2$OH		3.76, dd (10.4, 4.5)	3.61, dd (10.5, 5.5)	3.72, dd (11.7, 4.2)		3.72, dd (11.7, 4.2)
		3.61, dd (10.4, 9.0)	3.56, dd (10.4, 5.3)	3.63, dd (11.7, 3.3)		3.62, dd (11.7, 3.3)

* In order to compare with the sub-structural unit in **4**, the NMR data are parallel listed with No. 1′ to 13′.

Table 2. ^{13}C NMR (DMSO-d_6, 150 MHz) data of compounds **1–6**.

No.	1	2	3	4	5	No.	2	3	4	6 *
1	168.1	168.3	168.1	170.8	168.1	1′	168.2	168.1	166.1	170.9
2	118.5	119.1	118.7	118.6	118.5	2′	118.8	118.6	117.3	118.6
3	153.2	152.4	153.0	153.1	153.1	3′	152.8	153.1	154.6	153.1
4	129.6	129.7	129.6	129.6	129.6	4′	129.6	129.6	129.5	129.6
5	141.3	140.8	141.2	141.2	141.3	5′	141.1	141.2	142.3	141.3
6	48.9	48.8	48.9	48.6	48.8	6′	48.8	48.8	48.9	48.8
7	30.8	30.7	30.7	30.6	30.8	7′	30.9	30.8	30.8	30.7
8	36.8	36.9	36.9	36.6	36.8	8′	36.4	36.7	36.5	36.6
9	26.8	26.9	26.9	26.3	26.8	9′	26.6	26.7	26.5	26.5
10	25.9	25.0	25.0	25.4	25.8	10′	25.7	26.0	25.2	25.2
11	35.0	32.6	32.6	32.2	35.7	11′	35.8	34.9	33.6	33.6
12	69.3	71.6	71.3	70.9	69.5	12′	69.3	69.3	74.7	74.8
13	55.4	53.3	53.3	52.9	56.0	13′	56.2	55.7	58.8	58.8
14	174.9	174.4	174.1	173.9	175.2	14′	173.3	172.7	168.1	168.1
3-CH$_3$	19.5	19.5	19.5	19.5	19.5	3′-CH$_3$	19.5	19.5	19.6	19.5
5-CH$_3$	18.6	18.5	18.5	18.5	18.6	5′-CH$_3$	18.5	18.6	18.6	18.6
7-CH$_3$	19.7	19.8	19.7	19.7	19.7	7′-CH$_3$	19.7	19.7	19.7	19.7
13-CH$_2$OH	60.1	60.6	60.6	60.1	61.3	13′-CH$_2$OH	61.5	59.9	56.7	56.7

* In order to compare with the sub-structural unit in **4**, the NMR data are parallel listed with No. 1′ to 13′.

2.2. Antifungal Activity

P. theae (HQ832793) and *C. gloeosporioides* (HQ832797) were isolated from foliar lesions of tea leaf and their pathogenicity to tea leaf were verified both in vitro and in vivo.

The antifungal activity of the isolated compounds was evaluated by the paper disc inhibition assay and the minimum inhibitory concentration (MIC) was determined by the paper disc dilution method as described previously [15]. The new compound **4**, containing a β-lactone ring, exhibited moderate inhibitory activities with MIC of 50 μg/disc for *P. theae*, while showing little activity with *C. gloeosporioides*. Hymeglusin (**6**), a typical β-lactone antibiotic and equisetin (**7**), a terpenoid alkaloid, exhibited potent inhibitory activities against both fungi with an MIC value of 25 μg/disc.

2.3. Toxicity Evaluation

All the isolated compounds were evaluated for their anti-proliferative effects on zebrafish embryos (Figure 4). Equisetin (**7**) showed strong anti-proliferative effects, leading to embryo death with an EC$_{50}$ value of 0.12 μM at 48 h after treatment; mortality at 96 h was 100% even at the lowest concentration of 0.625 μM. All the alkenoic acid derivatives (**1–6**) exhibited much lower toxicity on zebrafish embryos. At 48 h, most of the zebrafish embryos were alive when treated with compounds **1–6** even at the highest concentration of 10 μM. At 96 h, *bis*-alkenoic acid ester-type compounds (**2–4**) and the β-lactone-type compound (**6**) showed anti-proliferative effects on zebrafish embryos at 10 μM.

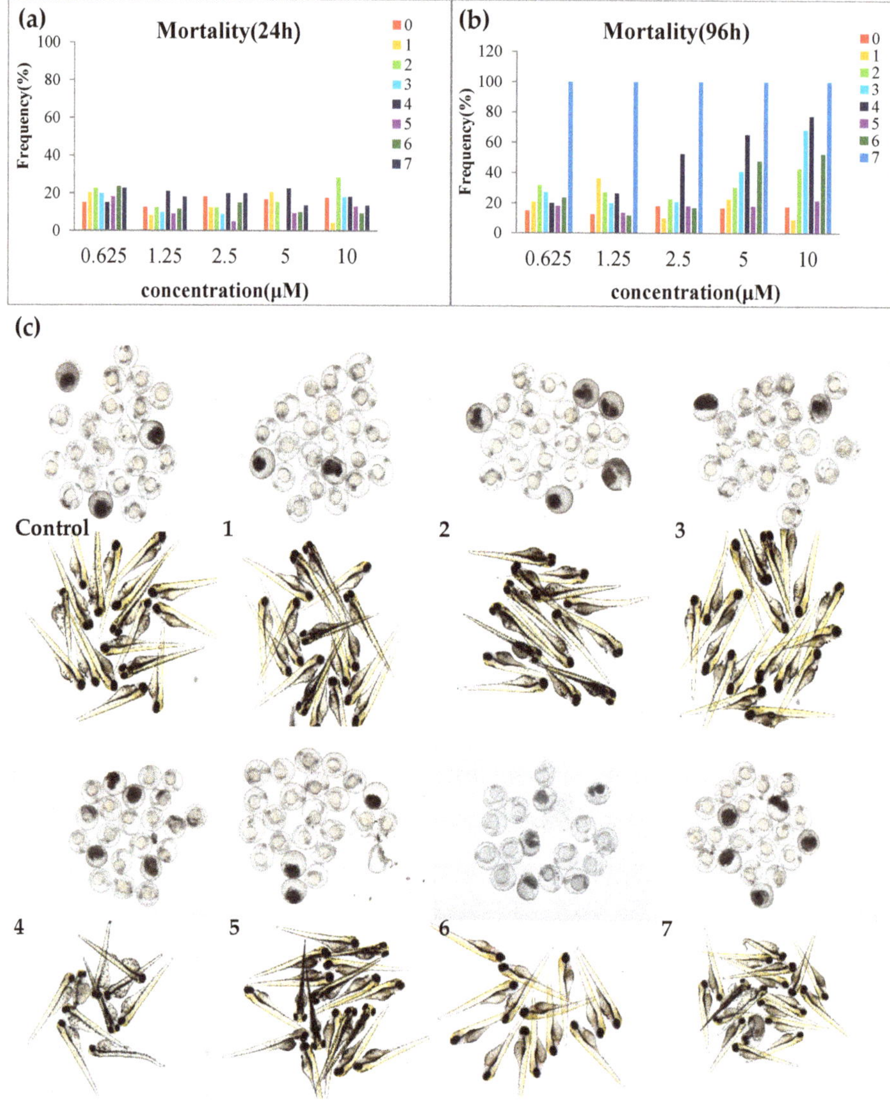

Figure 4. Toxicity evaluation of compounds **1–7** on zebrafish model. (**a**) Mortality of zebrafish embryo at 24 h. (**b**) Mortality of zebrafish larva at 96 h. (**c**) Photograph of zebrafish embryo (24 h) and zebrafish larva (96 h). The final concentration of compounds **1–6** was 10 μM and that of compound **7** was 0.31 μM.

3. Discussion

In this study, four new compounds, including a new alkenoic acid together with three new *bis*-alkenoic acid esters were isolated from the ethyl acetate extract of marine-derived fungus *F. solani* H915. Chemically, the relative configuration of these compounds was confirmed by their NOESY spectra; the absolute configuration was revealed by product elucidation via alkaline hydrolysis experiments. All the compounds were evaluated for antifungal activity. The new compound **4**, containing a β-lactone ring, exhibited moderate inhibitory activities with MIC of 50 μg/disc for *P. theae*. Hymeglusin (**6**) and equisetin (**7**) exhibited potent inhibitory activities against both fungi with an MIC of 25 µg/disc. As two new isolated tea pathogenic fungi, typical antifungal drugs, such as fluconazole and fluorocytosine did not show inhibitory activity on them, even at a concentration of 50 µg/disc. The β-lactone ring containing alkenoic acid derivatives possibly prevents fungal diseases in tea plants.

However, the toxicity of the antifungal regents is important. Although showed potent activity, equisetin exhibited strong anti-proliferative effects on zebrafish embryos and larvae, indicating high toxicity. Alkenoic acid derivatives showed much lower toxicity to zebrafish. Thus, hymeglusin (**6**), an alkenoic acid derivative with a β-lactone ring from *F. solani* H915, is a low-toxicity, potent antifungal agent against tea pathogenic fungi.

4. Materials and Methods

4.1. General Experimental Procedures

An electrospray ionization source (ESI)-equipped Q-Exactive Mass spectrometer (Thermo Fisher Scientific Corporation, Waltham, MA, USA) was used to analyze the HR-ESI-MS data. A Shimadzu UV-260 spectrometer (Shimadzu Corporation, Tokyo, Japan) and a Perkin-Elmer 683 infrared spectrometer (PerkinElmer, Inc., Waltham, MA, USA) were used to obtain the UV and IR spectra, respectively. A JASCO P-200 polarimeter (JASCO Corporation, Tokyo, Japan) with a 5 cm cell was applied to measure the optical rotation value. The NMR spectra with TMS as the internal standard were taken on a Brucker Avance III 600 FT NMR spectrometer (Bruker Corporation, Billerica, MA, USA).

4.2. Fungal Strain and Fermentation

The strain *Fusarium* sp. H915 was isolated from mangrove sediments at the Zhangjiangkou Mangrove National Nature Reserve, Fujian, China, suing the tablet pour method. The internal transcribed spaces (ITS) region was amplified and sequenced using the general primers ITS1 and ITS4. The ITS region of the fungi is a 576 bp DNA sequence (GenBank Accession Number: KY978583) that showed 99% identity to *F. solani*. The strain was deposited at the China Center for Type Culture Collection (CCTCC) as accession number M2017150 and Marine Culture Collection of China (MCCC) as accession number MCCC 3A00957. The fungus grew well on the rice medium in artificial sea water. Carbohydrate fermentation was conducted by subculturing the fungus in rice medium in artificial sea water and incubated at 28 °C for 30 days at a standing position.

4.3. Extraction and Isolation

The rice medium (10 kg) of *F. solani* H915 was extracted with ethyl acetate (20 L) trice and concentrated under reduced pressure at 40 °C to yield 16.4 g of the residue.

The EtOAc extract (15 g) was fractionated over a column packed with silica gel (300 g, Yantai Chemical Industry Research Institute, Yantai, China), eluted with petroleum ether-ethyl acetate (v/v) (20:1; 10:1; 5:1; 2:1; 1:1; each 1.0 L) and chloroform-methyl alcohol (v/v) (50:1; 20:1; 10:1; 5;1; 2:1; 0:1; each 1.0 L), to afford 10 fractions (Fr. 1–10). Further separation was conducted on the fractions with antifungal activity (Fr. 6 and 8) and the high-yielded fraction (Fr. 9). Fr. 6 (4.6 g) was separated over a Cosmosil reversed-phase C18 (100 g, 75 µm, Nakalai Tesque Co. Ltd., Kyoto, Japan) column and eluted with CH_3OH/H_2O (10–100%, each 0.5 L) to provide nine subfractions (Fr. 6-1–Fr. 6-9). Fr. 6-7 (1.3 g) was purified by a preparative Cosmosil octadecylsilane (ODS) column (250 mm × 20.0 mm inner diameter (i.d.), 5 µm, Cosmosil, Nakalai Tesque Co. Ltd., Kyoto, Japan), isocratically eluted with acetonitrile-H_2O (42:58, v/v) to obtain compound **6** (300 mg, 1.8% yield). Fr. 6-9 (1.3 g) was separated by the preparative ODS column and eluted with acetonitrile-H_2O (65:35, v/v) to obtain compound **7** (57 mg, 0.38% yield). Fr. 8 (1.5 g) was also fractionated by an ODS column and eluted with CH_3OH/H_2O in gradient mode and 10 subfractions were obtained (Fr. 8-1–Fr. 8-10). Fr. 8-9 (270 mg) was purified by preparative HPLC column and eluted with acetonitrile-H_2O (65:35, v/v) to obtain compound **4** (30 mg, 0.18% yield). Fr. 9 (3.8 g) was also separated over an ODS open-column and eluted with CH_3OH/H_2O to yield nine subfractions (Fr. 9-1–Fr. 9-9). Fr. 9-7 (325 mg) was then isolated by preparative ODS column and eluted with acetonitrile-H_2O (32:68, *v/v*) to yield compound **1** (19 mg, 0.11% yield) and compound **5** (14 mg, 0.085% yield). Preparative HPLC purification of Fr. 9-9 (290 mg), eluted with acetonitrile-H_2O (67:33, *v/v*), led to the isolation of compound **2** (10 mg, 0.061% yield) and compound **3** (62 mg, 0.38% yield).

Fusaridioic acid A (**1**): white amorphous powder; $[\alpha]_D^{25}$ value was $-6°$ (*c* = 0.1, CH_3OH); IR (KBr) (ν_{max}): 2925, 2355, 1683, 1593, 1253, 1189 and 1079 cm^{-1}; UV (MeOH) λ_{max} (log ε): 201 (3.70), 230 (3.43) and 270 (3.90) nm. ^1H NMR (600 MHz, DMSO-d_6) and ^{13}C NMR (150 MHz, DMSO-d_6) spectral data were listed in Tables 1 and 2; HR-ESI-MS: *m/z* 341.1968 $[M - H^+]^-$ (calcd. for 341.1964 $C_{18}H_{29}O_6$).

Fusariumester A_1 (**2**): colorless viscous oil; $[\alpha]_D^{25}$ value was $+2°$ (*c* = 0.1, CH_3OH); IR (KBr) (ν_{max}): 3400, 2930, 2864, 2354, 1695, 1632, 1379, 1250 and 1174 cm^{-1}; UV (MeOH) λ_{max} (log ε): 203 (3.69), 231 (3.47) and 265 (3.74) nm. ^1H NMR (600 MHz, DMSO-d_6) and ^{13}C NMR (150 MHz, DMSO-d_6) spectral data were listed in Tables 1 and 2; HR-ESI-MS: *m/z* 665.3883 $[M - H^+]^-$ (calcd. for 665.3901 $C_{36}H_{57}O_{11}$).

Fusariumester A_2 (**3**): colorless viscous oil; $[\alpha]_D^{25}$ value was close to $0°$ (*c* = 0.1, CH_3OH); IR (KBr) (ν_{max}): 2928, 2857, 2361, 2341, 1698, 1603, 1247 and 1175 cm^{-1}; UV (MeOH) λ_{max} (log ε): 201 (4.17), 233 (3.80) and 270 (4.19) nm. ^1H NMR (600 MHz, DMSO-d_6) and ^{13}C NMR (150 MHz, DMSO-d_6) spectral data were listed in Tables 1 and 2; HR-ESI-MS: *m/z* 665.3882 $[M - H^+]^-$ (calcd. for 665.3901 $C_{36}H_{57}O_{11}$).

Fusariumester B (**4**): white powder; $[\alpha]_D^{25}$ value was $-12°$ (*c* = 0.1, CH_3OH); IR (KBr) (ν_{max}): 2927, 2661, 2361, 2341, 1823, 1715, 1698, 1616, 1234, 1149 and 1046 cm^{-1}; UV (MeOH) λ_{max} (log ε): 202 (4.06), 232 (3.80) and 272 (4.23) nm. ^1H NMR (600 MHz, DMSO-d_6) and ^{13}C NMR (150 MHz, DMSO-d_6) spectral data were listed in Tables 1 and 2. HR-ESI-MS: *m/z* 647.3798 $[M - H^+]^-$ (calcd. for 647.3873 $C_{36}H_{55}O_{10}$).

4.4. Alkaline Hydrolysis of **2–4** and **6**

Each compound (1 mg) was dissolved in a mixture of 5% KOH-dioxane (1:1, 4 mL). The solution was stirred at room temperature overnight. The reaction mixture was neutralized with 1% HCl. The neutralized solution was filtered through a 0.22 μM filter membrane to afford the test solution. Each test solution (20 μL) was analyzed over an Cosmosil C18 column (250 mm × 4.6 mm i.d., 5 μm, Nakalai Tesque Co., Ltd., Kyoto Japan) and eluted with water (A) and acetonitrile (B) at a flow rate of 1.0 mL·min^{-1}, according to the following gradient program: A from 60% to 40% and B from 40% to 60% during 0–30 min. The detection wavelength was 254 nm. The alkaline hydrolysis products were identified by comparison of their retention times (t_R) with those of compounds **1** and **5**.

4.5. Antifungal Activity

The antifungal activity against tea pathogenic fungi *P. theae* and *C. gloeosporioides* was evaluated according to a previously described method [10,11]. Petri plates were used in the test. A piece of tested fungal strains cylinder agar with diameter of 0.6 cm was placed in the center and sterile blank paper discs (0.5 cm diameter) were placed 2 cm from the growing mycelial colony. Approximately 20 μg of each compound was loaded to each paper disc. DMSO was used as the blank control. The plates were incubated at 28 °C until mycelial growth enveloped the control discs. Then 10 μL of each compound's DMSO diluted solution ranging from 0.3125 to 10 μg/μL was added onto paper discs. The inoculated plates with impregnated paper discs were incubated at 28 °C for seven days. The lowest concentration of active compound that could inhibit visible mold growth was recorded as the MIC. The experiment was repeated and recorded thrice.

4.6. Antiproliferative Effects on Zebrafish Embryo

Compounds **1–7** were dissolved in DMSO at a concentration of 10 mM and stored at -20 °C until analysis. Toxic activity of the isolated compounds was analyzed with the anti-proliferative effects on zebrafish *Danio rerio* embryo, according to previously described methods [15]. 3,4-Dichloroaniline was used as a positive control. Fertilized embryos were collected following natural spawning in 12-well plates. Embryos were periodically checked for death and developmental delay. At 6 h post-fertilization (HPF), embryos were arranged in 12-well plates at 20 embryos/well. Pure compounds were then added to the desired concentration DMSO as the vehicle control. DMSO was kept at 0.5% of the total volume. Embryos were grown at 28 °C and examined with a Leica stereomicroscope at 24, 48, 72 and 96 h post treatment. The death rate was recorded every. All animal procedures were conducted in accordance with all appropriate regulatory standards under protocol P18030102 (approval date: 2018-03-01) approved by the Xiamen University Institutional Animal Care and Use Committee.

Supplementary Materials: The following are available online at http://www.mdpi.com/1660-3397/16/12/483/s1, Figure S1–S8: Spectra of compound **1**, Figure S9–S17: Sepctra of compound **2**, Figure S18–S26: Spectra of compound **3**, Figure S27–S34: Spectra of compound **4**, Figure S35: HPLC chromatograms of compounds **1–6** and alkaline hydrolysis products of **2–4**, **6**.

Author Contributions: X.-X.T. supplied and identified the bacteria. S.-Z.L. performed the antifungal activity assays. Y.-K.Q. and J.-G.L. supervised the project. X.Y. isolated the compounds.

Funding: The National Natural Science Foundation of China (grant number 81773600) supported this study. The project was also supported by the National Basic Research Program of China (973 Program) (No. 2015CB755901), the International S&T Cooperation Program of China (No. 2015DFA20500) and Xiamen Ocean Economic Innovation and Development Demonstration Project (16PZP001SF16). Scientific Research Foundation of Third Institute of Oceanography, SOA. (No.2016002, 2017002), Xiamen Science and Technology Program (Nos.3502Z20172009 and 3502Z20182029)

Acknowledgments: The authors thank to Mei-Juan Fang and Zhen Wu for their kind help on structural elucidation. Thanks to Ming-Yu Li for his help in zebrafish toxicity evaluation.

Conflicts of Interest: The authors have no conflict of interest to declare. The founding sponsors had no role in the design of the study; in the collection, analyses, or interpretation of data; in the writing of the manuscript and in the decision to publish the results.

References

1. Feller, I.C.; Lovelock, C.E.; Berger, U.; Mckee, K.L.; Joye, S.B.; Ball, M.C. Biocomplexity in Mangrove Ecosystems. *Annu. Rev. Mar. Sci.* **2010**, *2*, 395. [CrossRef] [PubMed]
2. Spellberg, B.; Bartlett, J.G.; Gilbert, D.N. The future of antibiotics and resistance. *N. Engl. J. Med.* **2013**, *368*, 299–302. [CrossRef] [PubMed]
3. Saravanakumar, D.; Vijayakumar, C.; Kumar, N.; Samiyappan, R. PGPR-induced defense responses in the tea plant against blister blight disease. *Crop Protect.* **2007**, *26*, 556–565. [CrossRef]
4. Saha, D.; Dasgupta, S.; Saha, A. Antifungal activity of some plant extracts against fungal pathogens of tea (*Camellia sinensis*). *Pharm. Biol.* **2005**, *43*, 87–91. [CrossRef]
5. Gunasekera, T.; Paul, N.; Ayres, P. The effects of ultraviolet-B (UV-B: 290–320 nm) radiation on blister blight disease of tea (*Camellia sinensis*). *Plant Pathol.* **1997**, *46*, 179–185. [CrossRef]
6. Ponmurugan, P.; Baby, U.; Rajkumar, R. Growth, photosynthetic and biochemical responses of tea cultivars infected with various diseases. *Photosynthetica* **2007**, *45*, 143–146. [CrossRef]
7. Sanjay, R.; Ponmurugan, P.; Baby, U. Evaluation of fungicides and biocontrol agents against grey blight disease of tea in the field. *Crop Protect.* **2008**, *27*, 689–694. [CrossRef]
8. Sarkar, S.; Ajay, D.; Pradeepa, N.; Balamurugan, A.; Premkumar, R. Evaluation of chemical and neem pesticides against *Pestalotiopsis theae* causing grey blight disease of tea. *Ann. Plant Protect. Sci.* **2009**, *17*, 252–253.
9. Chakraborty, B.; Basu, P.; Das, R.; Saha, A.; Chakraborty, U. Detection of cross reactive antigens between *Pestalotiopsis theae* and tea leaves and their cellular location. *Ann. Appl. Biol.* **1995**, *127*, 11–21. [CrossRef]
10. Staeubert, C.; Krakowsky, R.; Bhuiyan, H.; Witek, B.; Lindahl, A.; Broom, O.; Nordstroem, A. Increased lanosterol turnover: A metabolic burden for daunorubicin-resistant leukemia cells. *Med. Oncol.* **2016**, *33*, 1–10. [CrossRef] [PubMed]
11. Li, W.; Tang, X.X.; Yan, X.; Wu, Z.; Yi, Z.W.; Fang, M.J.; Su, X.; Qiu, Y.K. A new macrolactin antibiotic from deep sea-derived bacteria Bacillus subtilis B5. *Nat. Prod. Res.* **2016**, *30*, 2777–2782. [CrossRef]
12. Aldridge, D.C.; Giles, D.; Turner, W.B. Antibiotic 1233A, a fungal β-lactone. *J. Chem. Soc. C* **1971**, 3888–3891. [CrossRef]

13. Kumagai, H.; Tomoda, H.; Omura, S. Biosynthesis of antibiotic 1233A (F-244) and preparation of [14C]1233A. *J. Antibiot.* **1992**, *45*, 563–567. [CrossRef] [PubMed]

14. Ratnaweera, P.B.; de Silva, E.D.; Williams, D.E.; Andersen, R.J. Antimicrobial activities of endophytic fungi obtained from the arid zone invasive plant *Opuntia dillenii* and the isolation of equisetin, from endophytic *Fusarium* sp. *BMC Complement. Altern. Med.* **2015**, *15*, 220. [CrossRef]

15. Parng, C.; Seng, W.L.; Semino, C.; McGrath, P. Zebrafish: A preclinical model for drug screening. *Assay Drug Dev. Technol.* **2002**, *1*, 41–48. [CrossRef] [PubMed]

MDPI
St. Alban-Anlage 66
4052 Basel
Switzerland
Tel. +41 61 683 77 34
Fax +41 61 302 89 18
www.mdpi.com

Marine Drugs Editorial Office
E-mail: marinedrugs@mdpi.com
www.mdpi.com/journal/marinedrugs